D0055799

TOXIN
TOXOUT

Also by Bruce Lourie and Rick Smith

Slow Death by Rubber Duck:
The Secret Danger of Everyday Things

TOXIN TOXOUT

GETTING HARMFUL CHEMICALS OUT OF OUR BODIES AND OUR WORLD

BRUCE LOURIE / RICK SMITH

St. Martin's Press New York

Printed in the United States of America. For information, address St. Martin's Press, 175 Fifth Avenue, New York, N.Y. 10010.

www.stmartins.com

Library of Congress Cataloging-in-Publication Data

Lourie, Bruce.

Toxin toxout : getting harmful chemicals out of our bodies and our world / Bruce Lourie, Rick Smith. — First U.S. edition.

p. cm.

Includes bibliographical references and index.

ISBN 978-1-250-05133-2 (hardcover)

ISBN 978-1-4668-5586-1 (e-book)

1. Detoxification (Health) 2. Environmental health. 3. Environmentally induced diseases—Nutritional aspects. I. Smith, Rick, 1968– II. Title. III. Title: Toxin toxout.

RA784.5.L68 2014

613—dc23

2013049688

St. Martin's Press books may be purchased for educational, business, or promotional use. For information on bulk purchases, please contact Macmillan Corporate and Premium Sales Department at 1-800-221-7945, extension 5442, or write specialmarkets@macmillan.com.

First published in Canada by Knopf Canada, a division of Random House of Canada Limited, Toronto

First U.S. Edition: May 2014

10 9 8 7 6 5 4 3 2 1

For Biz and Jen

CONTENTS

FOREWORD

BY FLORENCE WILLIAMS,

AUTHOR OF *Breasts: A Natural and Unnatural History*

WHEN I READ BRUCE LOURIE and Rick Smith's first book, *Slow Death by Rubber Duck*, I was first hooked by the best book title of the year and then won over by the authors' originality, humour and passion. At the time I was working on my own book, about the natural history of breasts, and I, like so many readers, was fascinated and horrified by all the industrial chemicals coursing through our bloodstreams and body fat. How did they get there and what did it mean for our health?

In *Slow Death*, Lourie and Smith experimented on themselves, seeing if they could influence their exposures by changing their habits, diets and environments, such as by sitting in a room with a stinky carpet freshly sprayed with a stain-resistant coating. These off-gassing chemicals, like so many others found in everyday products, foods and materials, end up in our bodies, where they have the potential to build up and interfere with cell signalling and other biological processes. Some may act like hormones or damage DNA. We've entered a new frontier of science, and not enough people or regulatory agencies are paying attention. But thanks to books like *Slow Death*—and now *Toxin Toxout*—this is starting to change.

A dozen years ago, the promise of health lay in genomics. Scientists decoded our genes, thinking they would unlock the secrets of disease. But it turns out that most of our genes are pretty excellent all by themselves. It's when the outside world communicates with our genes, changing how they're expressed, that things become interesting. And the outside world is more complicated than ever before. It's now believed that a great number of diseases—everything from autism to allergies to cancer—are caused by genes and the environment acting together. Just as we once mapped the genome, scientists are now calling for a way to map our "exposome," beginning in the womb. Where do we live and what do we eat and smoke and how high are the radiation levels?

Now that we are more alert to the uninvited molecular guests in our bodies, we increasingly want answers: how do we reduce our exposures to the most worrisome substances and how do we get them out once they're there? People ask me these questions all the time. The reduction part is a little easier. If people want to take precautions, they can try to avoid microwaving plastics, for example, or avoid using heavily scented personal care products. But the toxout part is harder. From studies of people and marine mammals, we know that one of the best-known methods for removing toxic chemicals from your body is to deliver them straight to your infant through breastfeeding, but that's hardly comforting (and by the way, it's my duty as a breast devotee to tell you that the benefits of breastfeeding still outweigh the risks). But wouldn't it be nice to get those chemicals out of there before the baby shows up?

I think most people would agree: The best solution is not to pollute ourselves in the first place; for example, by replacing risky substances with safer ones whenever possible in industrial processes. Thanks to governments and manufacturers in North America, we've already come a long way toward reducing our exposures to lead, benzene, asbestos, organochlorine pesticides, tobacco, radon and many other baddies. But there are plenty left to feel uneasy about.

So I'm grateful for *Toxin Toxout*, the important latest installment of Lourie and Smith's work. With this book they have embarked on a fearless detox mission, wading through the often woo-woo world of cleanses, saunas, filters and chelation to see what, if anything, helps. They smelled some nasty stuff, got some needle pokes and Lourie even passed out. Detox isn't easy. Thanks, guys, we owe you one!

Florence Williams
Washington, D.C.
July 2013

ACKNOWLEDGEMENTS

THEY SAY THAT THE SECOND BOOK, album or movie is the most difficult. But what the heck do "they" know anyway? What about *Led Zeppelin II*, *The Empire Strikes Back* and *The Two Towers*, we thought to ourselves as we toiled to ensure that this second toxic tome was as well turned as the first. If we've succeeded, it's because of the many people who gave so generously of their time to this complicated project. If we didn't, it's entirely our fault.

We want to begin by thanking the extraordinary Random House of Canada team of Louise Dennys, Marion Garner, Amanda Lewis, Brad Martin, Matthew Sibiga and our editor, Paul Taunton, for their enthusiasm and incredible efforts as they brought this volume to life. Our agent, Rick Broadhead, made sure that the book project went smoothly; we even read our contract. Again, Kathryn Dean has corrected our grammar and punctuation, making us sound articulate.

Our employers, past and present, have always been co-conspirators in the best sense. At Environmental Defence, Canada's premier pollution fighters, thanks go to Bob Davies, David Donnelly, Aaron Freeman, Aviva Friedman, Stephanie Kohls, Gillian McEachern, Maggie MacDonald, Matt Price, Alanna Scott, Eric Stevenson, Deborah Sun de la Cruz and Sarah Winterton. Millions of Canadians continue to benefit every day from Environmental Defence's groundbreaking *Toxic Nation* and *Just Beautiful* campaigns. Thanks to the members of the *Just Beautiful* campaign cabinet, especially Jennifer Ivey, Sarah Harmer, Donna Bishop, Lisa Borden (she of uptown herbal tea fame), Gillian Deacon, Wendy Franks, Sarah Jay,

Trish McMaster, Brian Phillips, Joanna Runciman, Nicole Rycroft, Ersilia Serafini, Laurie Simmonds, Tracie Wagman and Dr. Shirley Zabol. The Ivey Foundation not only supports the work of numerous outstanding environmental initiatives in Canada, but also provides Bruce with the encouragement and freedom to pursue book writing and all that it entails, and for that, we are deeply grateful. Thanks to Ed Broadbent and the Broadbent Institute board and staff for recognizing that the economy of the future will be green and that we need to create jobs through reducing all types of pollution. This project would not have happened without the unflappable good humour and organizational moxie of Rachel Potter, who coordinated the complicated experimental logistics and kept us moving forward. There were some days when we couldn't even see her desk behind the piled boxes of incoming and outgoing experimental vials.

We would also like to gratefully acknowledge the generous support of the Catherine Donnelly Foundation, the J.W. McConnell Family Foundation and the Global Greengrants Fund for making this book possible.

We are appreciative of the staff at the Division of Biological Sciences of the University of Missouri; Chemir Analytical Services (Maryland Heights, Missouri); Pacific Toxicology Laboratories (Chatsworth, California); Solstas Lab Partners (Chicago, Illinois); and AXYS Analytical Services (Sidney, British Columbia) for their diligent analyses. To all the people we interviewed, consulted, pestered and relied upon along the way, *mille mercis*: Eugenia Akuete, Jennifer Arce, Lisa Archer, Laura Batcha, Beto Bedolfe, Judi Beerling, Chuck Benbrook, Janine Benyus, The Big Carrot—especially Heather Barclay, Maureen Kirkpatrick, Daiva Kryzanauskas and Carol Roche, Linda Birnbaum, Jessa Blades, Ted Boadway, Carl Gustav Bornehag, Jim Brophy, Antonia Calafat, Ray Civello, Marc Cohen, Terry Collins, Ken Cook, Shannon Coombs, Cindy Coutts, Julie Daniluk, Philippa Darbre, Mia Davis, Miriam Diamond, Katherine DiMatteo, Sarah Elton, Maria Emmer-Aanes, James Ewles, Alex Formuzis, Jeff Gearhart, Stephen Genuis,

Linda Gilbert, Richard Grace, Michael Green, Rebecca Hamilton, Scott Hickie, Gary Hirschberg, Matt Holmes, Jane Houlihan, Stephen Huddart, Mauro Iacoboni, Markus Koenig, Annie Leonard, Andrew Leu, David Love, Alex Lu, Stacey Malkan, Mónica Marín, Ron McCormick, Jason McLennan, Beth McMahon, José Mestre, Carl Minchew, Clarissa Morawski, Thomas Mueller, Pete Myers, Liza Oates, Siobhan O'Connor, Erik Olson, Jenny Pape, Mike Partain, Sam Pedroza, Michael Perley, Greg Potter, Horst Rechelbacher, Noah Sachs, Amarjit Sahota, Marshall Stackman, Rena Steinzor, Shanna Swan, Julia Taylor, Betsy Thomson, Curt Valva, Jasper van Brakel, Adria Vasil, Fred Vom Saal, John Warner, Charles Weschler, Heather White, Bill Whyte, Bryce Wylde and Tom Zoeller. We owe a particular debt to those families who donated their time to our organics experiment and to Dr. Peter Erickson, Rodney Palmer and Peter Sullivan, who went far beyond what is reasonable in terms of their generosity with their time, their homes and their professional services.

Our families remain our sources of inspiration. Bruce is incredibly thankful that his beautiful girls, Biz Agnew, Ellen and Claire, are even more enthusiastic about "the book" than he is and that they've been so tolerant of his having to spend so much time on it. And after 50-plus years of life, he appreciates the privilege of his own upbringing like never before—particularly, the love and support of his parents, Allan and Grace. Were it not for Jennifer Story's love of fresh food and organic and local ingredients, Rick would doubtless still be puzzling over how to cook Kraft Dinner—and eating way too much of it. She is his best friend and partner in all things. Rick's fantastic sons, Zack and Owain, and his nephews, Gabriel and Noah Smith-Vaz, are the kids he has in mind whenever he's writing: They deserve to live in a world that does much, much better at making children's health a top priority. Rick's grandfather John Braive died during the writing of this volume. The extent to which he took his citizenship seriously should be a model for us all.

Finally, we thank you: the tens of thousands of people who continue to participate in the global success of *Slow Death by Rubber Duck*, whether through hearing us speak, buying a copy of the book for your grandchildren or simply "liking" us on Facebook. This book is first and foremost dedicated to your curiosity and commitment and to the many questions that provoked the investigations in the pages that follow.

INTRODUCTION

You live your life like a canary in a coalmine
You get so dizzy even walkin' in a straight line

—THE POLICE, "CANARY IN A COALMINE,"
Zenyatta Mondatta, 1980

"HOW DO I GET THIS STUFF out of me?"

The question—half concern, half exasperation—came from a balding, middle-aged man in the audience at the Sydney Writers' Festival in 2010. We were there to talk about our book *Slow Death by Rubber Duck*, which went on to become an Australian bestseller. To illustrate our point about toxic chemicals in consumer products, we had stopped at a local supermarket and come to our author session at the festival armed with a number of the products we'd written about: baby bottles containing the hormone-disrupting chemical bisphenol A (BPA), toothpaste with the thyroid toxin triclosan and numerous kitchen implements slathered with noxious non-stick coatings. There we sat on the festival stage, the fruits of our shopping arrayed around us and the audience staring with newfound horror at the so-familiar, so-surprisingly-toxic icons of our global consumer culture. As we travelled across Australia, Canada, the United States and Europe to promote *Slow Death by Rubber Duck,* this scene repeated itself over and over again. Wherever we went, our shopping turned up the same toxic consumer products. And in every city, interested crowds gathered to hear about the book and the results of our self-experimentation. We recounted how just a few days of using certain consumer products more than doubled our personal mercury content and increased our bodies' BPA levels by over 7 times, our phthalate levels by 22 times and our triclosan levels by nearly 3,000 times.

Though we would love to claim prescience, we'd be lying if we didn't admit to just a little surprise at the global appeal of *Slow Death by Rubber Duck*. It turns out that even more than we'd imagined, the threat of toxic chemicals is an international concern. In this age where disparate parts of the globe are bound together like never before—where permutations of obscure economic indicators in Europe can send shock waves through markets everywhere and funny YouTube videos are seen simultaneously by hundreds of millions of people—the entire industrialized world covets the same brands, shops in similar stores and is exposed to the same harmful synthetic chemicals. The new pollution affects us all. We have become, all together, the proverbial canaries in the coal mine.

In the wake of the intense interest in *Slow Death,* we began planning this second book. Right off the bat, the need to focus on detox was obvious. After all, every audience and every interviewer quizzed us about the hazards of toxic chemicals, how they became common in our everyday lives and how they are linked to human disease. And all of these people, including the middle-aged man in Sydney, the hippie TV journalist in Stockholm, the shock jock in Chicago and the naturopathic doctor in Calgary, were preoccupied with the same question: How *do* we get this stuff out of our bodies? If this wasn't convincing enough, at every speaking event we attended, at least one person, and often several, spoke to us about their favourite detox treatments: saunas, diets, potions and pills. They asked us whether we'd used any of these detox therapies to rid ourselves of our own, well-documented toxic body burdens. (As it turned out, we hadn't.)

None of this reaction was surprising. Once you find out that one unwelcome by-product of our modern age is that pollutants are indiscriminately taking up residence in your body, you would obviously want to know what to do about it. And you'd want answers to these questions: How *can* we reduce our toxic chemical intake and how *can* we rid our bodies, our lives and ultimately our economy of these synthetic hazards? Finding some honest and specific responses became the purpose of *Toxin Toxout.*

Quick Refresher

Before we press forward, let's take a step back for a second and ask the question, Why should people be concerned about the synthetic chemicals that surround us? It's simple: The scientific evidence linking these chemicals to human disease has become even more convincing than it was when *Slow Death* was released. Exposure is widespread. A recent study by the Environmental Working Group (EWG) in the United States found 232 toxic chemicals in the umbilical cord blood of 10 babies from racial and ethnic minority groups. Since 1995, EWG's body burden testing has found 553 different industrial chemicals, pollutants and pesticides in 149 Americans across 27 different states.[1] In Canada, Environmental Defence released a report in the summer of 2013 that involved testing the umbilical cord blood of three newborns for the presence of 310 different synthetic chemicals. In total, 137 different chemicals (including things like DDT, PCBs and flame retardants) were detected in the three newborns. The report was the first published Canadian data of this kind, and the startling results demonstrated that Canadian children are born pre-polluted.[2] To date, less research has been conducted in developing countries, but we can reasonably assume that to the extent people are exposed to the same chemicals in those areas, the results will be similar.[3]

The verdict on individual chemicals is in. For instance, BPA has now been linked with elevated risk of heart disease, infertility and diabetes-like effects, among others.[4] The presence of triclosan, that ubiquitous "antibacterial" ingredient in everything from personal-care products to footwear, is increasing dramatically in the bodies of people and in lakes and rivers, and has now been linked to increasing rates of allergies.[5] The evidence of harm from mercury exposure continues to accumulate, with recent experiments revealing effects as varied as autoimmune diseases in adult women and hormone alterations in kids.[6]

And there's news from Parkersburg, West Virginia, the destination of a road trip that Bruce wrote about in *Slow Death*. Recall that the citizens in Parkersburg brought forward a class action lawsuit against Dupont, the manufacturer of perfluorooctanoic acid (PFOA), commonly known as C8 and used to make Teflon and other non-stick and stain-resistant coatings. This was done after it was revealed that the company's Parkersburg plant had been releasing C8 since the 1950s, contaminating the local drinking water. As part of the settlement agreement of the lawsuit, an independent panel of public health experts was established to determine whether there was a probable link between C8 exposure and disease in the town. This panel is known as the C8 Science Panel, and the investigation is one of the largest studies of its kind in history, with seventy thousand local residents having donated blood samples which were then tested for contaminants and compared with individual health records. The panel issued its final Probable Link report in late 2012 and has concluded that there is a probable link between C8 and pregnancy-induced hypertension, preeclampsia, testicular cancer, kidney cancer, ulcerative colitis, thyroid disease and medically diagnosed high cholesterol.[7]

The results of being exposed to one synthetic chemical are worrisome enough, but the mixture of toxic chemicals in our bodies, *all at once*, has a more profound and complex effect than the chemical industry would like us to believe. The twisted logic of numerous chemical industry moguls goes like this: Let's look at individual chemicals in isolation and try to set some magical "safe" level for each one. Like a light switch, if the level of this one chemical in your body is below the "safe" level, you'll be good to go, the light won't come on, there will be no effect. If you're above the level, whoops!

This rationalization falls apart, of course, when we have hundreds and thousands of synthetic chemicals in us all together. When it comes to the health effects triggered by these pollutants, one plus one equals more than two: These chemicals can actually *amplify* each other's individual impacts.[8]

Significantly, an increasing number of studies are now indicating that the extent to which people can withstand the toxic chemical cocktail we are all exposed to is highly variable and at least partly based on their genetic makeup.[9] But do you want to play that kind of Russian roulette?

A Kind of Progress

As we noted in the introduction to *Slow Death*, the speed of the debate surrounding environment and health is exciting, and as a consequence of strong recent scientific evidence linking toxic chemicals to serious human disease, there has been a marked and positive change in the public's everyday behaviour. As one example, the organic food and beverage industry has grown rapidly worldwide. In 2010 the global market for certified organic food and drink was estimated to be US$59.1 billion, which represents a 9.2 percent increase over the $54.1 billion in sales in 2009.[10] Traditionally, the organic food industry was based mainly on fresh produce, and while organic fruits and vegetables retain the highest sales growth, the industry has expanded into many processed food products.[11]

Other trends reflect an increasing desire on the part of consumers to avoid toxins in everyday life. In the cleaning products aisle, Method and Seventh Generation now compete for market share with the likes of Clorox and Procter & Gamble. That has led to the surest sign of market success—namely, the big companies getting into the game with products like Clorox's Green Works. And Martha Stewart's line of cleaning products is made with non-toxic ingredients: surely a cultural bellwether if ever there was one.

Eco-friendly products and "green consumerism" are not a new trend, but they're consistently an area where opportunities exist for creative brands and entrepreneurs to respond to changing consumer needs. Over the last 10 years, Trendwatching.com, one of the world's leading trend firms, has repeatedly listed "Eco" as one of the top consumer trends to watch for.[12] Further, in a 2011 global survey of executives from commercial companies around the world, 70 percent of respondents had placed sustainability permanently on their management agendas, all within the last six years.[13] Over two-thirds of respondents say their organization's dedication to sustainability has increased and will continue to do so.[14]

As a result of the accumulating science and consumer awareness in this area, governments are acting. Not quickly. But they are moving. The laws governing BPA present a good example of this progress. Following the Canadian ban on BPA in baby bottles in 2008 (Canada was the first country in the world to do this), the European Union followed suit in 2010, and in early 2011, China did as well. Effective in 2013, France has further outlawed the use of BPA in plastic food containers. And though progress is slow on a federal level in the United States, many states have now followed Canada's lead: BPA bans relating to various plastic baby products are now in place in 12 states. Notably, in April 2013, the state of California added BPA to its Proposition 65 list—a list of toxic chemicals that cause cancer or birth defects. This could result in BPA showing up on the warning labels of not only baby bottles, but hundreds of other household products as well. Thanks to the work of Environmental Defence Canada and others, the Canadian government is considering classifying triclosan as "toxic" under the country's pollution law, and both the Canadian and American Medical Associations have called for restrictions on the household use of this chemical.[15] When governments get their acts together and ban or restrict a substance, really important and dramatic things can happen. The graphs in Figures 1 to 3 tell the tale.

Figure 1. Decreasing DDT levels in West Germans between 1972 and 1995

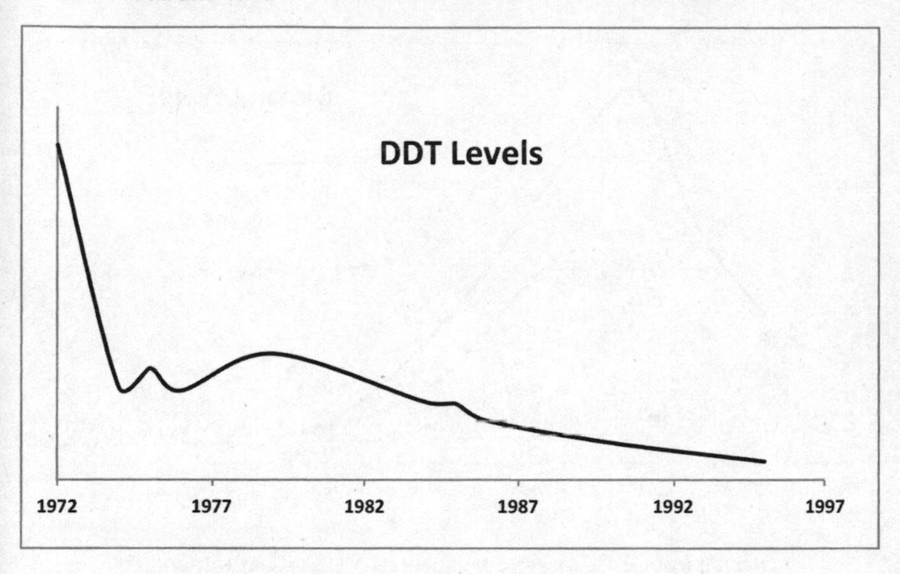

Adapted from D. Smith, "Worldwide Trends in DDT Levels in Human Milk," *International Journal of Epidemiology* 28 (1999): 184.

Figure 2. Decreasing dioxin levels in Americans born in 1950, 1970 and 1980

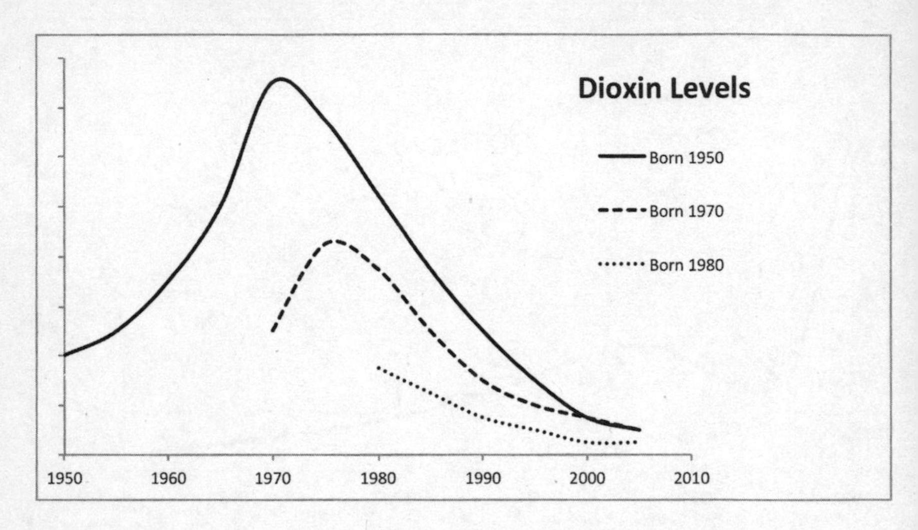

Adapted from P. Pinsky and M. Lorber, "A Model to Evaluate Past Exposure to 2,3,7,8-TCDD," *Journal of Exposure Analysis and Environmental Epidemiology* 8 (1998): 325.

Figure 3. Decreasing lead levels in Canadians

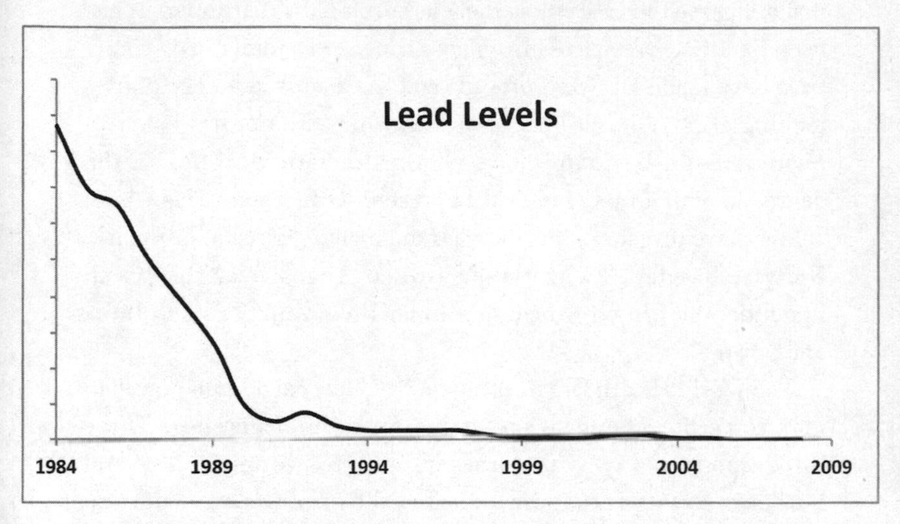

Adapted from Health Canada, *Risk Management Strategy for Lead,* February 2013: 25, accessed April 19, 2013, http://www.hc-sc.gc.ca/ewh-semt/pubs/contaminants/prms_lead-psgr_plomb/index-eng.php.

DDT is one of many nasty pollutants (in this case, a pesticide) that are persistent, carcinogenic and hormonally disruptive. It has been banned or restricted in 57 countries, and a total of 102 countries have made DDT imports illegal.[16] As a consequence, studies looking at DDT levels in breast milk—like the one reflected in Figure 1 from Germany—have shown dramatic declines in the degree to which this chemical is present. Other countries where studies have revealed a downward trend include Canada, Denmark, Norway, Sweden, Switzerland, Turkey, Yugoslavia, the Czech Republic, the United Kingdom, China (Hong Kong), Israel, India and Japan.[17]

The World Health Organization (WHO) has used human milk to monitor other chemical pollutants for several decades. Three WHO-sponsored,13-year studies of milk from women in several European countries examined dioxins, dibenzofurans and dioxin-like PCBs and showed a general downward trend in levels of exposure to these chemicals.[18] Another study compared the body burden data of a type of dioxin in Americans born in 1950, 1970 and 1980, respectively, and showed a dramatic decrease in levels of this chemical over time (see Figure 2).[19] These are all the positive results of measures taken over the last 20 years to reduce emissions that create dioxins, PCBs and dibenzofurans.

Finally, lead exposure in Canada (Figure 3) and other countries throughout the world has decreased substantially since the early 1970s, mainly because leaded gasoline and lead-based paints were phased out and the use of lead solder in food cans was virtually eliminated.[20] The consumer and regulatory trends we've just described illustrate one theme related to the toxic chemical issue that will recur throughout this book: *Only when people exercise their power as both citizens and consumers will there be solutions to the problems caused by damaging chemicals in the environment.* As citizens we must demand that our governments respect the health and future of all by properly restricting and managing unsafe chemicals. As consumers we need to protect ourselves and our families by making informed choices, given the lack of corporate concern for our health. It is through firing on both these cylinders that a greener future will be brought about.

Detox. For Real.

Let's get back to the here and now. So while governments are making modest progress in getting toxins out of our bodies in the longer term, we need to ask ourselves this question: Can we do anything to reduce pollutants in our bodies, and in our kids' bodies, in the short term? Answer: Yes. There are things that can be done right now. And that is what *Toxin Toxout* is all about.

Based on our self-experimentation and interviews with experts, we examine, in the rest of this book, how toxins enter our bodies through absorbing, eating, breathing and drinking—and how to keep them out. With the help of some friends in the beauty industry, we compare body paraben and phthalate levels when green versus conventional cosmetics products are used. Then we move on to some groundbreaking experiments with kids to find out how organic versus conventional diets can affect their body pesticide levels.

We go on to explore the multibillion-dollar detox industry through the stories of several fascinating individuals who have devoted years of their lives to the complex, expensive and sometimes painful journey of personal detoxification. Then we take a deep dive into learning how the human body naturally removes toxic chemicals and whether we can enhance those natural detox mechanisms. Getting up close and personal, we set out to see whether some of the better-known detox treatments actually work, so in a reversal of the tox-"IN" testing of *Slow Death* and the other chapters in this book, we try some tox-"OUT" self-testing. Bruce travels to California for an electromagnetic-free sleep and to Texas for an intravenous detox experience called "chelation therapy." And after spending hours in an infrared sauna back in Canada collecting sweat, we discover that detox treatments aren't necessarily for the faint of heart. Do these alternative therapies work? This is the big question, and we find the answers the hard way. Our next experiment takes us to Michigan, where Rick spends an entire day in a parked car to check out the "new car smell."

With our experiment and personal detox results in hand, we step back to take a look at the bigger picture with an investigation of the burgeoning field of "green chemistry." The journey begins with Bruce climbing North America's biggest toxic trash pile to see where broken toys end up and to discover ways to avoid the creation of toxic waste in the first place. We also look at some fascinating toxin-free emerging technologies and plumb the depths of the challenges faced by those who are trying to create a greener, less toxic world.

The final chapter of the book summarizes what we have learned, within a new framework for understanding, about how to detox our bodies, our lives and the economy. Combining new research results with our detailed investigation of personal detox strategies and our understanding of the potential new green economy future, we create the "Toxin Toxout Top 10" list.

In a sea of publications about toxic chemicals and detox regimes, this book stands out because we demonstrate through *doing*. And we aren't hustling the latest fad diet—in fact, quite the opposite: We provide a prudent appraisal of the sometimes confusing world of detox. Our opinions on what works and what doesn't are based on direct experimentation, in addition to extensive research. In many cases, the experiments have never been done before. What you're reading here is a first!

Here we go.

ONE: WELLNESS REVOLUTION

~ Rick lathers up ~

A woman who doesn't wear perfume has no future.

—COCO CHANEL

THE SONG SAYS, OF NEW YORK CITY, that if you can make it there, you can make it anywhere.

So I think it's an auspicious sign that the Sustainable Cosmetics Summit is now an annual event in the Big Apple. After arriving at the midtown Manhattan hotel where the event was being held in May 2012 and squeezing by Jude Law's largish camera crew (he was filming a scene in the hotel lobby), I found myself in a fluorescent-lit room with a couple of hundred people. Despite the beckoning sunshine outside, they were—as evidenced by the large coffee mugs in hand—digging in for two days of back-to-back seminars.

"Eclectic" would be an apt word to describe the crowd. Over the next few days, I met representatives of wild shea nut harvesters from sub-Saharan Africa, jojoba farmers and producers of obscure natural scents and oils. Buttoned-up, corporate types from companies like Colgate-Palmolive mingled with former (or current) patchouli-redolent hippies and über-trendy, black-clad, celebrity hairdressers. What brought this diverse group together?

The smell, literally, of success.

Amarjit Sahota is the summit organizer and the head of London-based Organic Monitor, a research and consulting company specializing in the organic sector. According to Sahota, the growth of the natural and organic personal-care market is outpacing all other segments of the cosmetics industry. And we're talking here about an industry that's way more than makeup. From shampoo to shaving products, from fragrance to facial cleanser, these people define pretty much everything in your bathroom that you don't put in your mouth. Though still comprising less than 10 percent of industry sales in Europe and North America, natural and organic cosmetics were nonetheless worth US$9.1 billion in 2011 and are growing at almost US$1 billion per year. I asked Sahota what was behind this phenomenon.

"It's really a manifestation of a broader trend," he told me. "What we are seeing globally is a general rise in ethical consumerism. People have become more discerning about the products that they buy. They're asking more questions: Where is it made? Does it have synthetic chemicals? Is it being produced in an ecological way? This is happening with food, cosmetics, household cleaning products, clothing, toys, and the list goes on."

Linda Gilbert, a Florida-based pollster who since 1990 has been tracking public attitudes related to sustainability and who presented her findings to the summit, agreed with Sahota. "The consumer interest in sustainability, and in particular limiting their exposure to pollution and toxic chemicals in their everyday lives, is a tidal wave. It's transformational," she told me. "It is impacting virtually every industry that you look at. The home-building industry, with the drive to low-VOC building materials. The food industry, with increased scrutiny of packaging. The transportation industry. The garden product industry. Consumers are becoming more and more aware of ways they can avoid chemicals in their lifestyles."

Her blunt conclusion? "If companies don't re-tool and re-invent themselves, they're not gonna be in business." Gilbert used the example of bisphenol A in baby bottles to illustrate her point. Over the past few years, the growing consumer backlash against the hormone-disrupting chemical has ensured that non-BPA bottles now dominate the market.

Gilbert calls this consumer awakening a *wellness revolution*. "Consumers in the United States have this growing sense that there's a personal environment that they can and should control," she explained. "They can't control the global environment, but they can control what goes in their home, what goes on their lawn, what they're breathing in because of fragrances and body lotion." She maintains that for health reasons, and often with no particular reference to environmental concerns, consumers are looking to reduce exposure to synthetic chemicals in their daily life. Though about 15 percent of the population will never be interested in these issues, Gilbert admits (what she calls the "drill baby drill" demographic), her measurements indicate that the appetite for sustainable cosmetics is growing: "86 percent of all Americans are interested in making greener choices in their purchasing and, more specifically, 41 percent would do so with personal-care products in 2012, up three points since 2010. They are ready to make changes for a more eco-friendly lifestyle, and—significantly—willing to change brands to accomplish this."[1]

Perhaps no person is more steeped in the phenomenon of green consumerism than Toronto author and columnist Adria Vasil. For 10 years, week in and week out, Vasil has been writing her cheeky "Ecoholic" column in *Now* magazine, answering questions about all manner of green conundrums ("Which blouse is more environmentally friendly: the one made from organic cotton or the one that's rayon from sustainably harvested trees?") and has parlayed her position as the "Dear Abby" of green into three bestselling books. Though some "green products" haven't done so well, she told me as we chatted in my local park after the summit, she sees continual growth in public concern about toxic chemicals in personal-care products. "Pre-recession, when everyone was jumping on the green bandwagon and consumers en masse were starting to prioritize the environment as their number one concern, you were seeing everyone from bra stores to mainstream furniture outlets suddenly advertising a green line," she said. "A lot of those came and went within a year of their release because they just didn't sell."

"What? Bra stores?" I asked, quite certain that I'd misunderstood.

"Sure," she laughed. "La Senza had a green line and an organic line. But they were really boring products. I mean, they were brown, and every other product in La Senza was tropically coloured!"

In addition to the questionable aesthetics of the ill-fated eco-bras, Vasil told me, she's convinced that the failure of some alleged "green" products and the success of others is related to the tangibility of the threat the products purport to address. "At the end of the day, the eco-bra was just a bra. And it tanked. But increasing demand for greener body-care products is far from plateauing," she said. "Every time I walk into a drugstore, I see a new product, a new mainstream brand pushing a green claim. In those areas where people intuitively understand the products have an impact on their own health or their kids' health, there's some lasting power."

When we realize the pollution is personal, we're motivated to pay to avoid it.

In the Beginning . . .

If there's one company in the beauty industry that has made natural ingredients the very core of its brand, it's Aveda (its slogan is "The Art and Science of Pure Flower and Plant Essences"). Founded by Austrian celebrity hairdresser Horst Rechelbacher in 1978, Aveda was subsequently sold to Estée Lauder in 1997 and has since grown into a global presence. Like Apple, which has tried to define itself with a certain green grooviness, Aveda took some flak for not paying adequate attention to its environmental standards. But it has made up for lost time, now banning phthalates, parabens, sodium lauryl sulphate (SLS) and a variety of other nasties from its formulations. (See Table 1 for a short primer on these, and some other common chemicals found in cosmetics.) Aveda also boasts that it is the first beauty company to run its manufacturing on 100 percent wind power, and it uses all post-consumer materials in its packaging.

Table 1. A few problem chemicals

The Chemical	What It "Does"	The Problem
Triclosan	Synthetic antibacterial agent used as a preservative in things like shaving creams, hair conditioners, deodorants, liquid soaps, hand soaps, facial cleansers and disinfectants Also used in some products that don't make antibacterial claims	Animal research studies have shown it to be an endocrine disruptor at low levels of exposure Concerns about increased antibacterial resistance and resulting dangerous by-products like chloroform
DEP (diethyl phthalate) and DnBP (di-n-butyl phthalate)	DEP and DnBP are the most common phthalates in personal-care products—used for their ability to hold colour, denature alcohol, act as a lubricant and fix fragrance Found in heavily fragranced products like shampoos, lotions and perfumes	Known endocrine disruptors Adverse health effects from exposure include asthma and increased allergenicity, reproductive and metabolic disorders, and developmental and behavioural problems
Methyl paraben	Most commonly used preservative in cosmetics and personal-care products Methyl paraben is most often found in deodorants, shampoos, lotions and creams Also often used as unlisted "fragrance" ingredient	Found to mimic estrogen, which can lead to increased risk of breast cancer Can affect male reproductive functions Immune system and organ toxicants, allergens linked to dermatitis and have been shown to cause cancer in animals

SLS (sodium lauryl sulphate)	Petroleum-based chemical added to personal-care products as surfactants and foaming agents Commonly found in shampoos, shower gels, cleansers and toothpaste	Links to skin and eye irritation, organ toxicity, developmental and reproductive toxicity, and endocrine disruption The manufacturing process contaminates SLS with 1,4-dioxane, a probable human carcinogen

In Canada, Ray Civello has long been Mr. Aveda. He was the first to bring the line to Canada (starting by offering it in his salon over a muffin shop in Toronto's east end), and he still has exclusive rights. In the mid-1980s, Aveda's burgeoning delivery system was a bit quirky, and the product would regularly arrive frozen from its Minnesota distribution centre. Civello persevered, building double-digit sales growth, year over year, for more than two decades. A thoughtful environmentalist, opinionated about toxic chemicals and the need for more natural products, Civello has long put his money where his mouth is by supporting various charities such as Environmental Defence Canada.

I remember the first time I met Civello, nearly 10 years ago, at an event at Aveda's Canadian headquarters in suburban Toronto. Given that "fashionable" isn't the typical adjective that springs to mind when you picture environmentalists, my colleague and I were feeling just a tiny bit out of our depth at this gathering of nattily attired beauty-industry professionals. Civello was the epitome of cool in his finely tailored Prada suit, and as I listened to his motivational speech to his staff, it struck me how eloquent he was concerning the links between beauty, wellness and the environment. He later told me that his connection to Aveda and its social mission arose out of a difficult period in his life.

"I contracted mono, lost a lot of weight and was burned out on the whole rock-and-roll hairdresser lifestyle," he said. "It was because I got sick that I started reading about sustainability and wellness. I just couldn't function, and I realized I needed to change." In the late 1980s, environmental sensibility in the industry was pretty thin, he recalled. "Basically, hairdressers didn't care. In general, awareness of sustainability amongst people in beauty and fashion was not high. People didn't think about what was in the products they were using. The priority was always how to get the end result. It didn't matter about the product as much as the process," he said, throwing up his hands at the memory. "In fact, the more difficult and unusual the process, the better they considered the experience to be."

Things are completely different today. "Consumers are concerned. Yes, people want to look younger, but there's a line in the sand now between doing it naturally or getting it immediately." Rather than a quick fix like cosmetic surgery, more people are now asking questions like "Can I do this in a more gentle way? Can I be more preventative?" Civello sees this change in Aveda salons every day.

Aveda's growth has been directly related to its ability to tap into the sustainability impulses of its customers. In fact, if there is a godfather of green-ness in the cosmetics world, it is surely the stylishly goateed, septuagenarian founder of the company, Horst Rechelbacher. If you research Rechelbacher, quite a few of the images you'll find will show him drinking and eating Aveda products: nothing like chugging your hairspray and noshing on your lipstick to make a point. Such was his devotion to extolling the gospel of non-toxic, edible, plant-derived ingredients that he earned the nickname "Jojoba Witness."

I was fortunate to be able to spend some time with Rechelbacher the evening before his keynote speech to the Sustainable Cosmetics Summit in New York City. His full-floor Lower Fifth Avenue penthouse (part of which used to be owned by the photographer Robert Mapplethorpe) boasts a distracting, nearly-360-degree view of the city. That night it felt a bit more crowded than it might otherwise be: Products from Rechelbacher's new company, Intelligent Nutrients (IN), were piled up in the living room, ready for a launch the next day. "One of the first full lines to be certified USDA Organic," he proudly told me. (Any product with this label meets the strict organic certification requirements for food set down by the U.S. Department of Agriculture.) "Cosmetics are food," Rechelbacher continued. "What we put on our body ends up in our body." After selling Aveda to Estée Lauder for a tidy sum (US$300 million, I later found out), the organic cosmetics giant decided he wanted to start again.

"Activism to me is show and tell, rather than just tell," he declared, "because if you don't do it yourself, people will ask why." Disgusted with the mainstream companies of the cosmetics industry that persist in using unsafe ingredients in their products, Rechelbacher has designed IN to be a showcase of what is possible. "The Holy Grail in cosmetics is supposed to be organic, but this is only as good as the person who makes it," he said. "Even if you give a chef organic ingredients, she can still make shitty food. Or you can have a chef who is an artist, who understands food chemistry and flavours and their combinations, and she can make wonderful dinners. Cosmetics are the same.

"You know when people say they put lead in lipstick?" Rechelbacher said, leaning forward for emphasis. "They don't put lead in lipstick; chemists are not *that* insane. But lead shows up during chemical processing of heavy metals. It also shows up by processing carbon, which is coal. Hair dyes are based on carbon technology. You need primary colours to do all colours. So blue comes from cobalt, yellow comes from sulphite. Those are the primary colours that are used in all colourings—from cosmetics to cars. I said to myself, 'This is toxic. Why do we use this shit? What are the alternatives?'" He jabbed the air with his finger. "Food! You know there's blue corn, there's cranberries, and the list goes on. I'm the first to make a food supplement that's a lipstick, a lip colour. There are no toxins, just nutritional benefits."

The average woman allegedly eats about four pounds of lipstick during her lifetime. If she were using IN's lip products, it would probably do her body good, given the nourishing ingredients detailed in the company's promotional materials: nutrient-dense oils of acai, rosehip and black cumin and soothing, antioxidant-rich waxes and butters. Rechelbacher's product descriptions read like menu listings of appetizers in a high-end restaurant, not something you'd find in a makeup bag.

Perhaps the best indication of the respect people have for Rechelbacher is that he took more than an hour for what was scheduled to be a 10-minute keynote speech at the summit the next morning, and nobody seemed to mind. As he exhorted the assembled in the room to be "bold" and urged them to connect with their raw ingredients because "restoration starts from the ground up—with seeds," I couldn't help thinking how far cosmetics had come. What could be more refined, more urban and more . . . well . . . *antiseptic* than the beauty industry? And yet here they were, a couple of hundred industry leaders nodding along as Rechelbacher urged them to get mud on their boots and dirt under their fingernails.

Somewhere, that diva of haute couture, Coco Chanel, was turning over in her grave.

Leaving the VW Vans Behind

Cosmetics companies love their creation myths. Though the industry has become huge, with recent estimates of global annual sales in excess of US$250 billion,[2] many of makeup's most recognized brands are rooted in the humble beginnings of their charismatic founders.

Of the major cosmetics firms, the largest is L'Oréal, which was started by Eugène Schueller in 1909 when he started mixing his own "safe hair dyes" and flogging them directly to Parisian hairdressers. The industry was developed in the United States shortly thereafter by hard-driving individuals whose names became synonymous with their products: Elizabeth Arden, Helena Rubinstein and Max Factor. Revlon was launched in 1932, when Charles Revson, his brother and a chemist named Charles Lachman pooled their meagre resources to create a new manufacturing process for nail enamel. Of its creator, the Estée Lauder company says that "Estée Lauder founded this company in 1946 armed with four products and an unshakeable belief: that every woman can be beautiful."[3] And famously, Coco Chanel overcame a brutally impoverished childhood and parlayed an early interest in hat design into one of the most enduring fashion houses and perfume brands in the world.

The same kind of entrepreneurial moxie was alive and well at the summit in New York City, with many representatives of the sustainable cosmetics companies I met feeling the wind at their backs and looking forward to taking on the industry's big players. My friend Mia Davis, formerly the Organizing Director at the Campaign for Safe Cosmetics and now Executive Vice-President of an up-and-coming sustainable cosmetics company called Beautycounter, epitomized this feisty spirit. "A lot of these larger companies were small start-ups once, right?" she said. "We're a small, scrappy start-up right now, but we have our eye on the prize. We want to be really big and shaking up the industry in short order." Davis's career trajectory, from activist campaigner to corporate player, is itself an indication of the maturation of the sustainable cosmetics phenomenon. "I really enjoyed my years of working as an advocate for market change and improved legislative policies, and still feel very closely aligned with the campaign and the activist movement," she said. "But I decided the time had come for me to play the same sport, but move to a different team." In starting a new company with a team of social entrepreneurs, Davis told me she saw herself continuing to educate the public about toxic ingredients—but at the same time shaking up a cosmetics industry desperately in need of innovation.

Time and again in New York I met companies that, until the last few years, were small mom-and-pop operations, and now (much to their own surprise) are doing multimillion-dollar international sales. Typical of this granola-to-riches tale is Tampa-based Aubrey Organics.

Curt Valva, Aubrey Organic's CEO, laughingly confirmed that his company's founders "were a bunch of hippies selling products out of the back of their cars" and proudly rhymed off the firm's long series of firsts in the cosmetics industry: first to list all of its ingredients (1967); first to formulate products with jojoba oil (1970); first to be certified as an organic processor (1984). The world is changing, Valva says. "Sustainable cosmetics are going mainstream, and that's one of the biggest challenges for us as a company." And he's not alone in trying to reconcile his company's growth with an ongoing commitment to making the world a better place.

Live (Toxin-) Free or Die

From his office in the woods of New Hampshire, Bill Whyte, the founder of W.S. Badger Company, told me the story of how his business went from obscurity to having the top-selling natural sunscreen in the United States and Canada. After getting out of the army and kicking around for a bit, Whyte finally settled down and started a family in the Granite State. He built his own house, gardened organically and was happily working as a carpenter.

"And then one winter," Whyte said, in what I got the feeling was an oft-recounted tale, "I had really cracked fingers from working outside. It was *awful*. They would split and bleed, so I needed to do something. I was lying in bed one night with olive-oil-soaked socks covered in plastic bags. And Katie [Whyte's wife] turned to me and said, 'You know, that's really pathetic. You can do better than that.'" The next morning, Bill went to the kitchen and started concocting various mixtures. He made a balm out of beeswax, olive oil and other ingredients, and it healed his fingers nicely. "So a lightbulb went on in my head," he said. "I can sell this to carpenters and other people who work with their hands." And thus was birthed Badger, with the sum total of the company being Whyte filling tins late into the night, listening to rock and roll, and delivering on weekends.

For a while after the launch of his Badger Balm, Whyte told me he resisted entreaties from people urging him to create new products.

The makers of Tabasco sauce served as his model ("You gotta love a company that's been doing the same good thing with a single product since 1870"). When he eventually succumbed to the siren call of product diversification, he quickly focused on sunscreens. "At the time, there weren't any mineral sunscreens," he told me. "Even the natural sunscreens were using chemicals as the active ingredient." Always up for a challenge, and experimenting with mixtures in his kitchen, Whyte wondered whether he could make a sunscreen that would be healthful. While researching, he learned that non-toxic zinc oxide was used for the old lifeguard sunscreens.

Whyte's timing was perfect. Little did he know that the Environmental Working Group (EWG), well known for evaluating toxic chemical levels in consumer products and in the bodies of Americans, was about to launch its first-ever "report card" on sunscreens, ranking them for effectiveness and safety.

Of all the consumer products that EWG could have focused on, there was a reason they put sunscreen in their crosshairs, according to Ken Cook, EWG's perennially energetic president. "The federal government had not done anything to require significant efficacy or safety testing or anything for sunscreens. They had some guidelines that had been pending for, like, 30 years. It was ridiculous," Cook grumbled when we spoke via telephone. "We're not talking about mascara here or shampoo—something that's designed to make you look nice or make you clean; we're talking about [products that can prevent] skin cancers. And . . . companies were producing products that didn't work and advertising them falsely as staying on all day and being waterproof. Then on top of all that, you get to the chemical issues." As a result of their years of research, Cook and EWG had developed significant concerns about the chemicals in sunscreens leading to allergic reactions and immune system issues and—ironically—posing an increased risk of cancer. Fed up with government inertia, they decided to publish their own ranking of sunscreens according to rigorous safety criteria.

The public and media attention was huge.

Bill Whyte was out of the country at the time, so he found out over the phone that Badger was ranked at the top of the list of safe sunscreens. When the EWG analysis was featured on *Good Morning America*, Badger's sales exploded overnight. The company's inventory, which Whyte figured would last for a year, was gone in less than a week. And sales have been going through the roof ever since. But in the beginning, there was one small problem.

"Bill, I don't know how to break this to you," I told him, "but my family and I started to use your stuff around that time, and it made my kids look like Casper the Friendly Ghost. It was thick and white as hell," I told him.

"It sure was!" Whyte laughed.

"And it was useless in water. They'd leave little zinc slicks behind them as they swam."

"Oh, my gosh. Well, I think the formulation has much improved in the last five years." I assured him that it had and that my family was fond enough of his product that we'd be taking some on our upcoming camping trip.

At the daycare where I drop my kids each morning in Toronto, there's a big bin marked "SUNSCREEN." With the summer of 2012 being so sunny and hot in the area, as it was throughout much of the continent, parents were urged to make sure that each of their kids had a full tube in the bin at all times so the teachers could apply it at will. As a measure of the growing popularity of mineral sunscreens, tubes of Badger and other great brands like Canada's Green Beaver now have pride of place alongside smelly coconut-scented conventional products. And they're gradually taking over.

Though I'll grant you that it's not the most scientific measurement in the world, I don't think the greasy bin lies. Mineral sunscreens are on a roll.

Label Decoding

As I learned at the summit, perhaps nowhere are natural cosmetics more popular than in Europe. Some of the largest companies offering green beauty products are based there. "Our company has been walking the walk since the 1920s," said the baby-faced Jasper van Brakel, North American CEO of Swiss-based Weleda. "Using biodynamic ingredients, using organic ingredients. Never using chemicals. We manufacture holistic medicines as well as skin care, body care and baby care products. The world has caught up to what this company is doing. And we love it. It's great."

Now with US$450 million in global sales, Weleda was founded nearly one hundred years ago as a herbal medicine laboratory by that most prototypical of hipsters, Rudolf Steiner. Around this same time, the Austrian philosopher and social reformer was pioneering his Waldorf education system and a new spiritual movement known as "anthroposophy": an attempt to meld science and mysticism. Van Brakel tells me that Weleda is still privately held by two foundations that exist to further Steiner's worldview. "They've owned the company forever, they're never going to sell it. And they allow us to continue to work out of a principle, rather than out of the primary objective to make money."

I asked him if he felt territorial heading the North American operations of a company that, it could be argued, invented the whole concept of plant-based personal-care products but now had to watch some of the larger industry players crowding into the same space. "It's a huge issue for us," he replied, echoing Aubrey Organics' Curt Valva's concern. "There are many companies out there, some of them marketing themselves as organic and natural, that have ingredients in their products that are not great at all. . . . There are many different shades of green." Increasingly, Weleda is advising large retailers that carry its products, like Target and CVS, to set up a special natural and organic section for products "that meet a certain standard."

What that standard is and how to encapsulate it in a digestible way for consumers is, van Brakel admitted, the $64,000 question.

Some things are easy to label. "Sugar-free" is pretty self-explanatory. As are the increasingly common "peanut-free" or "gluten-free" tags sought by allergy sufferers. What do you do, however, when you have a beauty product and you want to advertise it as more "natural" or "organic"? How do you distinguish its complicated, unpronounceable ingredients as being healthier and safer than the equally byzantine ingredient lists on conventional beauty products? Not an easy task, especially considering the illogical and highly variable laws that exist in different countries. As just one example, some jurisdictions do not require chemicals like phthalates to be disclosed on ingredient lists, while others, like parabens or sulphates, must be shown.

Amarjit Sahota (head of the London-based research and consulting company Organic Monitor) identified this kind of perplexity in his market update report at the summit in New York. "There's no question that consumer confusion is the biggest challenge," he said and went on to point out that the rise of what he called "pseudo-naturals" (products whose claims of "green-ness" or "natural-ness" are dubious at best) was contributing to the problem. Ironically, the attempted fix for this phenomenon—the creation of labelling schemes to validate the benefits of truly green products—was now itself adding to the murk in the minds of consumers.

"Let me give you an example from the U.K.," Sahota explained. "Let's say you're in an organic shop and you're looking at personal-care products. You've got some products labelled 'organic' from the U.K., but then there are also a couple of different products from France. They've got two labels: one for 'organic' and one for 'natural.' Another brand from Germany has similar products, but they have different logos, and then you have two from the U.S. with two different logos again. So in this shop there are perhaps 20 brands with upwards of 8 different 'organic' and 'natural' labels." The obvious effect, Sahota pointed out, is the creation of profound confusion. There are probably 50 separate "organic" and "natural" labels for healthy personal-care products.

"Ecoholic" column writer Adria Vasil has the difficult task of distinguishing between the relative merits of green products for her readers every week, and she agrees with Sahota that the terrain is terribly confusing. "It's really hard to tell what label was created by a bunch of cynical marketers sitting around a table sipping on their lattes and what's an actual genuine label that is administered by an impartial third party and stands for something good and useful."

Thankfully, there would appear to be some simplification on the horizon. By late 2015, a brand-new European standard, "Cosmos," will amalgamate and replace the pre-existing Ecocert, BDIH, Cosmébio (France), Soil Association (U.K.) and other national labels. Between Cosmos, the Natural Products Association certification and USDA Organic, most sustainable beauty products will be captured (see Table 2 for more details).

Table 2. Label decoder

Label	Labelling Organization	Description
NATURAL PRODUCTS ASSOCIATION CERTIFIED	Natural Products Association Certified	Product is at least 95 percent naturally (plant or mineral) derived, excluding water. Contains no phthalates, parabens, synthetic fragrance, chemical sunscreens, petrolatum, siloxanes or formaldehyde-releasing agents, to name a few. Does not mean products are organic.
Certified Natural Cosmetics BDIH	BDIH Certified Natural Cosmetics	German label indicating product is free of synthetic fragrances, ethoxylated raw materials, silicone-based and genetically modified ingredients, synthetic dyes and others. Will be phased out in 2015 to merge with COSMOS.
ECO CERT ®	EcoCert	French organization certifying both "natural" and "natural and organic." Both have to be 95 percent or more of natural origin. Bans synthetic scents and colours, petrochemicals and processes like nano-tech and genetic engineering. To be phased out by 2015 to merge with COSMOS.

Label	Labelling Organization	Description
"Organic"	No labelling organization	Without a third-party seal, this wording is meaningless.
"Made with Organic ingredients"	No labelling organization	This wording has been approved for use on products by certifiers such as the USDA when the product contains 70 to 94 percent certified-organic content.
USDA ORGANIC	USDA Organic	Considered the top seal on the market, since body-care brands that bear it follow the strict USDA Organic rules for food. Products are at least 95 percent certified organic.
COSMOS ORGANIC	COSMOS (coming in 2015)	Developed at the European level by BDIH (Germany), Cosmébio and EcoCert (France), ICEA (Italy) and Soil Association (U.K.). Will ensure a product is biodegradable, free of nano-tech, GMO and officially endangered raw materials. No PVC packaging, and ingredients must be 95 percent certified organic.

Adapted from Adria Vasil's *Ecoholic Body* (Vintage Canada, 2012)

Brazilian Blowout Blow-Up

Yes, labels are good. They help consumers make more informed choices. But what happens when companies are prepared to, shall we say, economize on the truth?

In 2011, a small advocacy group based in Oakland, California, called the Center for Environmental Health (CEH) became suspicious that many manufacturers were being less than honest regarding the levels of organic ingredients in their products. California law is very clear on this score: In order to use the word "organic," you need to ensure that your product contains at least 70 percent organic ingredients. The CEH launched a time-consuming but simple exercise: It collected dozens of allegedly "organic" products—shampoos and conditioners, lotions, deodorants, toothpastes, you name it—and examined the detailed ingredient lists on the labels. Many of the lists showed few or no organic ingredients. And since ingredients are listed in order of predominance, from major to minor, the CEH calculated that each of the products contained far less than 70 percent organics. Frustrated with what they considered a clear flouting of the law, the CEH sued. The goal: to bring the companies into compliance with California law—either by having them remove their improper organic claims or by ensuring that they increased the proportion of organic ingredients in their products.

"We want to encourage organics, plain and simple," says Michael Green, Executive Director of CEH. "You can't successfully encourage organics if people don't believe that things are what they say they are on the label. Because who is going to go out of their way to buy this stuff unless there's some truth in advertising?" The lawsuit had a quick effect, Green said proudly. "The vast majority of the companies have changed what they were doing. Most of them took the word 'organic' off the label of the products that we challenged them on, and some of them then created another line of products that they *could* put that label on." For now, CEH's lawsuit has helped reinforce the fact that the word "organic" has value and a legal meaning and can't be used simply to hoodwink consumers. Unfortunately, another recent, high-profile example of labelling gone amok hasn't ended on nearly as positive a note.

In September 2010, Jennifer Arce, an experienced hairdresser based near San Diego, decided to try a new hair-straightening product manufactured by GIB LLC, commonly known as Brazilian Blowout. "I specifically chose Brazilian Blowout because it was advertised as the only hair-straightening treatment that improved the health of the hair, caused no damage, had no harsh chemicals and, most importantly, was formaldehyde free." To get the hang of the product so that she could eventually use it on her clients, Arce arranged for her sister, also a co-worker, to do a Brazilian Blowout on her.

Arce remembers the exact day, the exact moment, because her life hasn't been the same since.

"Within minutes of her applying it to my hair, my eyes were burning, my throat was burning, my lungs were burning, and I was having a hard time breathing. My symptoms were escalating, and my sister was having all the same issues." Arce told me that even when the two women moved outside to complete the procedure in the fresh air, they continued to feel sick. That night, when she went home, she was lethargic and could barely swallow, and she developed a terrible migraine. Her puzzled doctor attributed her ongoing problems to possible chemical poisoning. Nearly two years later, she still can't believe what has happened to her. "In the days following the treatment, I was having a hard time doing little, simple, everyday tasks," she said. "I couldn't turn on a stove or an oven because the gas fumes coming out would make me sick. I couldn't use any cleaning products. I couldn't pump gas. I couldn't use hairspray on myself or on my clients—just being at work around all these chemicals was a struggle."

The penny finally dropped for Arce when she realized that some of her co-workers were getting sick too. "Many of them had been on and off antibiotics for months after exposure to the Brazilian Blowout chemicals, and we all had the same symptoms." She and her sister started researching and discovered that an increasing number of people were recording similar complaints online. Moved to action, the U.S. Occupational Safety and Health Administration (OSHA) began testing the product—including in Arce's salon—and discovered that far from being formaldehyde free, Brazilian Blowout contained up to 10 percent of this volatile, cancer-causing agent.[4]

With GIB, the makers of the products, protesting all the while that they were safe, the State of California sued. Arce was one of the hairdressers involved in the action. While the public debate raged and the lawyers were on the case, she was back at work, trying to make a living, having been forced to move salons when her previous boss refused to ban Brazilian Blowouts in the shop. The allure of the product was too much. "Oh, yeah," Arce said when I asked if the thing worked. "My hair never looked so good after using it. It was beautiful, shiny, you barely had to dry it. Some of the clients who do know it contains the chemical don't care, because it's making their hair so beautiful. And some of them will say, 'Beauty costs!'"

Arce now works at a new salon that refuses to do Brazilian Blowouts, but she and her co-workers have already been devastated by the experience in a number of ways. "A lot of us are on inhalers now. All of the diagnostic testing, sinus X-rays, chest X-rays, MRIs, CT scans, nose probes, breathing tests, EKGs—we've all taken many weeks off of work because we've been so ill from our exposure to these products." Arce suspects that some of her colleagues will be leaving the business entirely.

After appearing on the *Today* show and other media, telling the story of her harrowing experience, Arce now gets letters from hairdressers across the country with similar heart-breaking tales. One of the things they commiserate about is the incredible ending to the story. Even though GIB LLC was forced to pay a settlement of $600,000 to the State of California and "cease deceptive advertising that describes two of its popular products as formaldehyde-free and safe," Brazilian Blowout is still on sale.[5] While GIB is now obliged to identify formaldehyde in its labelling and instructions, the U.S. government lacks the power to force a product off the shelf if an obstreperous company is unwilling to do so voluntarily. Though it's now banned in some countries,[6] in the U.S., Brazilian Blowout has prevailed over its detractors and is still being used today in salons from coast to coast.

If you wanted one of the starkest examples of the abject failure of the U.S. regulatory system to ensure the safety of personal-care products, Brazilian Blowout would surely be it. But the Brazilian Blowout case is not an isolated one, and this is perhaps not surprising, since the U.S. law allegedly protecting

Americans from being poisoned by their bathroom products dates back to 1938—well before the creation of many modern synthetic chemicals.

When I caught up with Lisa Archer, the National Coordinator of the Campaign for Safe Cosmetics, I asked her how bad the current situation actually was. Fifteen minutes later, she'd rattled off a list of worrisome stories and statistics as long as my arm. "There are carcinogens in baby shampoo as well as phthalates and other problematic chemicals. Another example is mercury in skin-lightening creams, and there have actually been some cases of mercury poisoning from the use of these products."

In addition to lacking the legal authority to require recalls of damaging products, current U.S. (and Canadian) law allows companies to put dozens of secret ingredients in their products without disclosing them on the label. As long as a manufacturer can make the case that a particular synthetic chemical is a component of their fragrance formulation, they don't have to list it on their packaging. Thus, you'll never see the word "phthalates" on a product. In a recent Canadian study, Environmental Defence tested personal-care products and discovered, on average, 14 secret ingredients per product that were—quite legally—not disclosed on the ingredient lists.[7] The other reason that cosmetics safety is a concern in the U.S. is that the capacity of the U.S. government to oversee the industry is "virtually nonexistent and completely ineffective" according to Archer. "In a nutshell, FDA [the U.S. Food and Drug Administration] is pathetically understaffed and underfunded. We're talking about a budget of about $10–12 million per year and roughly 10 full-time staff to govern a $60 billion cosmetics industry. There's no way, given their current capacity and their current powers, that they could actually protect the public and ensure that cosmetics are safe."

In response to the concerns voiced by her organization and others, Archer was pleased that the U.S. Congress recently scheduled its first hearing on cosmetics safety in over 30 years.

In addition, a variety of Democratic representatives introduced a bill (the Safe Cosmetics Act of 2011), which would, if passed, result in a wholesale modernization of cosmetics regulation in the United States. But though there have been some victories at the state level (such as the California Safe Cosmetics Act that led to the prosecution of Brazilian Blowout), in the estimation of Ken Cook from the EWG, prospects of significant further statutory gains in the United States are "very grim, unfortunately."

Despite this, Archer and Cook are surprisingly upbeat regarding the pace of change in the United States. "There's a sort of a 'girlcott' going on, versus boycott," she told me. "Instead of *opposing* certain products, women are *supporting* companies who are more honest and transparent and using safer ingredients. And it's not just with cosmetics—you see it with BPA in baby bottles, you see organic food becoming more mainstream and things like that. And that's what I think is exciting. Even if the policy change is going to take a long time to happen, people are waking up to this issue. That market shift is going to continue to happen, driven by those conscious moms, in particular, who are changing their habits."[8]

Of Mennonites and Nail Salon Workers

It's an indicator of how far things have come that phthalates, surely the most unpronounceable word ever, have become such a poster child for consumer concern. Though there are many toxic chemicals that informed consumers are now on the lookout for on the labels of their beauty products—including sodium lauryl sulphate, siloxanes and Quaternium-15—phthalates are top of the list.

If you want to talk phthalates, the go-to expert is, without a doubt, Dr. Shanna Swan from Mount Sinai School of Medicine in New York City. Though I had interviewed her about the hormone-disrupting effects of phthalates for *Slow Death by Rubber Duck*, I had never actually met her. So when I sat down with this world-leading scientist amongst the dog walkers and pigeons on a sunny day in Madison Square Park, the first question I asked was whether we had learned much more about phthalates over the past five years.

"You know, it really depends on which phthalate you're talking about," she said. "Phthalates are a family of chemicals, and each one has a different toxicity, a different use, a different route of exposure."

She proceeded to run down her latest analyses, and the first study she mentioned involved some unique test subjects. "We did some testing recently of phthalates in the urine of Old Order Mennonites—10 pregnant women," she told me. "They had much lower levels of phthalates, BPA and triclosan in their bodies than the average American. One of the main reasons for this, we think, is that they don't use cosmetics. One woman had used hairspray, and she was the only one who had detectable levels of MEP [the breakdown product of DEP, a type of phthalate common in personal-care products]."[9] Another factor that Swan and her co-authors suspect accounted for the low levels in the volunteers was their general avoidance of cars and trucks. Phthalate levels in the interior air of cars can be elevated because of off-gassing from the upholstery, and this is particularly pronounced on warm days.[10] "Usually, Mennonites get around in horse-drawn buggies, but some of the women reported recently riding in a car or truck. We saw more MEHP [the breakdown product of the phthalate DEHP] in their urine." The third factor that Swan identified was the Mennonite habit of eating unprocessed foods, which they'd often grown themselves. Recent studies have shown markedly lower levels of BPA and some phthalates like MEHP when unprocessed food is consumed.[11]

"It's always DEHP that predominates in food. So why is that? Well, now I'm talking speculatively, but if you take a baby in the intensive care unit and feed it through a tube, you will measure DEHP in its urine. No question. A lot of it. Because the warm liquid pulls the DEHP out of the plastic in the feeding tube. I think this is probably what's happening with milk. Milking machines use a lot of plastic tubing. The DEHP from the plastic ends up in milk and cheese. It's fat soluble, so it accumulates in fatty foods. And so when people say, 'What can I do?' I say eat organic, unprocessed, fresh food. Your levels of DEHP will come down."

Another study that Swan mentioned—really the flip side of the Mennonite coin—is a startling look at phthalate levels in nail salon workers.[12] The levels of MBP [the breakdown product of the phthalate DBP] were significantly higher in manicurists after their work shifts. The use of gloves alleviated this problem, pointing to the nail products as the source of MBP. "Nail polish is bad, but perfume is the worst," said Swan, referring to recent studies demonstrating more uptake of phthalates from perfume than from any other personal-care product.[13] In the study, women who used perfume had three times the level of MEP in their urine as women who didn't wear perfume.

Given that phthalates surround us every day, virtually every human on the planet has the stuff coursing through their veins.[14] And to further compound the creepiness, elevated phthalate levels have been found in breast milk and umbilical cord blood, meaning that moms aren't just polluting themselves; they're passing their phthalate pollution on to their foetuses and nursing babies.

This is a problem because phthalates are hormone-disrupting chemicals. Once in our bodies, they are mistaken for estrogen and can create all the changes that estrogen achieves. Shanna Swan has long researched this phenomenon, including early publications about the possible role of phthalates in creating genital malformations of little boys.[15] A more recent study hints at neurobehavioural change resulting from exposure to phthalates.[16]

Hormones are like the traffic cops of our bodies. They tell everything—all critical processes—to "Stop," "Go" or "Slow Down." No wonder hormone mimickers like phthalates have such dramatic effects. (Tables 3 and 4 give a full rundown on what the most recent science is telling us about where we pick phthalates up in our daily lives and how they affect our bodies.)

Table 3. Where phthalates lurk[32, 33]

Chemical Name	Acronym	Source of Exposure
Diethyl phthalate	DEP	Used to bind cosmetics and fragrances; industrial uses include plasticizers, detergent bases and aerosol sprays
Dibutyl phthalate	DBP	Plastics such as polyvinyl chloride (PVC), adhesives, printing inks, sealants, grouting agents used in construction, additive to perfumes, deodorants, hair sprays, nail polish and insecticides
Butyl benzyl phthalate	BBP	Perfumes, hair sprays, adhesives and glues, automotive products, vinyl floor coverings
Di(2-ethylhexyl) phthalate	DEHP	Perfumes, flexible PVC products (shower curtains, garden hoses, diapers, food containers, plastic film for food packaging, bloodbags, catheters, gloves and other medical equipment)
Diisononyl phthalate	DiNP	Mostly in PVC as a plasticizer; remaining in rubber, inks, adhesives and sealants, paints and lacquers
Diisobutyl phthalate	DiBP	Nitrocellulose plastic, nail polish, explosive material, lacquer; application and properties similar to DBP: used in PVC, paints, printing inks and adhesives

Table 4. Recent science points to negative health effects of phthalates

Health Effects	Study Description	Year of Study
Asthma and increased allergenicity	Higher concentrations of phthalates BBzP, DEHP and DBP found in household dust in homes of children with asthma and allergies: association between exposure to phthalates and incidence of childhood allergy, asthma and related symptoms	2004, 2009, 2010, 2012[34]
Pregnancy loss	Higher concentrations of MEHP (DEHP metabolite) associated with higher occurrence of pregnancy loss	2012[35]
Metabolic disorders	Positive relationship between MEP (DEP metabolite), urinary concentrations and body size in overweight children	2012[36]
Reproductive disorders in male children	Positive relationship between DEP, DEHP, DBP, metabolites and reduced penile size, shorter anogenital distance or incomplete testicular descent	2005, 2006[37]
Reproductive disorders in adult males	Positive relationship between DEP, DEHP, DBP and increased sperm DNA damage, decreased sperm motility or decreased sperm concentration	2006, 2007[38]
Behavioural changes in male children	Prenatal exposure to DEHP and DBP may be associated with less male-typical play behaviour in boys (chemicals have the potential to alter androgen-responsive brain development)	2010[39]

And when it comes to beauty products, phthalates are only the tip of a diverse and complicated toxic iceberg.

Dangers of Deodorants (and Antiperspirants)

From the mountain of evidence linking phthalates to human health concerns, I turned next to another chemical that many alert consumers are now trying to avoid: parabens. Hormonally active chemicals that, like phthalates, mimic estrogen in the human body, parabens are added to countless consumer products—foods, pharmaceuticals and beauty products including antiperspirants and deodorants—as a preservative. Given their widespread use, it's not surprising that they're now found in the bodies of most people, including 95 percent of the American population.[17] Though parabens have not been scientifically scrutinized nearly as much as phthalates have, the examination of this preservative and its effects inside the human body is beginning to intensify. If there is one researcher in the world who can claim to have brought parabens into the public eye, it's Dr. Philippa Darbre of the University of Reading in England. In a widely cited 2004 study, Dr. Darbre and her colleagues found parabens in human breast tissue.[18] "There was a bit of a furor," Darbre told me over the phone from her laboratory. "Up to that point, it had been assumed that parabens, once they entered the human body, would be broken down by the liver. But something different altogether happens when they're applied directly to the skin: The parabens bypass the liver and remain intact."[19]

Darbre told me she was struck by the way that breast cancers often develop. "Between 50 and 60 percent of breast cancers start in the upper-outer quadrant of the breast, near the armpit." To explain why this is completely disproportionate and striking, she gave me a crash course in breast anatomy. "The breast is divided into seven regions: four quadrants, a central region, a nipple area and an axillary (or armpit) region. So if breast cancer started equally across all areas of the breast, we would expect to see less than

20 percent of cancers originating in each of those regions. But we don't, and 50 to 60 percent are up there in the upper-outer quadrant. Why? Is it because of all the chemicals being applied to that region?" This question has driven her to research the possible link between underarm products and breast cancer for over 15 years.

To further her 2004 study, Darbre continued to look for parabens in sample breast tissue from radical mastectomies. This time she used an even larger sample, and her suspicions were confirmed.[20] "Not only did we repeat our 2004 results, but we actually found even fourfold higher paraben levels in these samples." More interesting (and worrisome) in terms of implications for human health was the significant difference in levels of one paraben chemical in different parts of the breast. "Propyl paraben did seem to have a gradient, with more found in the axilla region than in the inner regions, which you might expect is coming from the underarm."

Do parabens cause breast cancer? As any cautious scientist would do, Darbre is quick to put her experimental results in context. "The fact that they're in the breast doesn't mean that there's necessarily a relationship with breast cancer. It's the first question. If they're not getting into the breast, then they can't have any effect on breast cancer." But even if they don't cause cancer, there may be other effects. Although cancer is the main concern, it actually represents only about 5 percent of clinical abnormalities of the human breast, with benign conditions such as breast cysts being the most common.

In response to the many women writing to her and complaining about painful breast cysts, Darbre has started looking into aluminum levels in breasts.[21] Here again she has found strong evidence linking antiperspirant use with disease. Aluminum is a common component of antiperspirants because it helps keep sweat off the wearer's skin by blocking sweat ducts. Breast cysts also occur when sweat ducts don't drain properly. Darbre has found strong evidence that, like parabens, aluminum levels are highest in the part of the breast near the armpit—also where a disproportionate amount of breast cysts are found—and that aluminum levels are higher in the fluid of breast cysts than in other parts of the body. Pretty convincing stuff.

What's the response of the chemical industry to Darbre's work? Although the presence of parabens and aluminum in breast tissue (specifically, in the part of the breast most likely to manifest disease) is now undeniable, the industry says that the levels of parabens are too minute to matter.[22] Even on this question, Darbre has tried to respond directly with further experimentation. Though manufacturers defend their *particular* parabens, it's not the fact that there is one type of paraben in breast tissue that's the problem, Darbe maintains; it's the fact that there are many. Parabens are a family of chemicals, and it's the effect of this potentially toxic and potent mixture that's the worry. In her most recent study, she took various parabens at the same concentrations she has measured in the human body and demonstrated that, in combination, they have an effect. "To my knowledge," she told me, "this is the first science suggesting that parabens have the capability of turning a normal breast cell into a transformed breast cell."[23] Transformed cells cannot be controlled by the body's normal processes and may be indicative of progression to a cancerous state. Scientists are producing a body of data leading to some valid concerns about parabens. But on the flip side, where is the data supporting claims that parabens are safe? As Darbre put it: "Where are the data showing that if you put parabens into all these things that get into people at all levels . . . that there are no effects? There aren't any such data. And my interpretation is that the current data imbalance is making companies nervous. Unfortunately, we are exposed to many chemicals each day that mimic estrogen and that have complementary action."[24]

In her own life, Darbre uses as few of these products as possible. And as Bruce illustrates in Chapter 4, there is growing evidence that sweating is actually an important mechanism of the body's detox system: The more you sweat, the more toxic chemicals you get rid of. Our over-air-conditioned, sweat-averse society's antiperspirant habit reduces our body's ability to clean itself while slathering on nasty pollutants. A double toxic whammy.

"A Whole New Ballgame"

I will confess that, in university, when my hipster friends were avoiding deodorant and rubbing crystals under their arms, I . . . was not. I value personal hygiene and a relative lack of stench from my fellow humans. Until I started writing this chapter, I still clung to my not infrequent use of Mitchum antiperspirant. Why? It works. As my grandmother always used to say, "Horses sweat, men perspire and women glow." I was going one further by trying not to perspire at all. With Toronto being 40 degrees Celsius for many days during the summer I was doing interviews and writing this book, I knew it would be hard to be taken seriously if I was perspiring through my shirt.

But there are now better ways to deal with the problem. Quite simply, sustainable cosmetics are more popular because products have become much better.

Just how much better was sketched out for me by Judi Beerling, Organic Monitor's Technical Research Manager. After working for over 30 years in the conventional cosmetics industry, Judi decided that the formulation of conventional cosmetics had become "a bit stagnant, with everybody doing the same sorts of things over and over." She was enticed to begin concocting formulations for sustainable cosmetics companies because of the intellectual challenge. She told me she now works out of a special lab she built for herself in her back garden.

When it comes to sustainable cosmetics, she told me during a coffee break at the New York Sustainable Cosmetics Summit, "it's a whole new ballgame." Five years ago, Beerling figured, you could really make only basic products. "Now you can make very elegant products that you would be hard pressed to distinguish from conventional cosmetics. Whether you can make them at the same cost, of course, is not so easy." As a rough approximation, Judi estimated that "85 to 90 percent of the ingredients and techniques [needed] to make sustainable cosmetics are now available, and this is increasing literally on a weekly basis. Now the challenge is to figure out how to combine them to get the best effect and to make the best cosmetics we can."

Hands down, the number one remaining challenge for formulating sustainable cosmetics, according to a number of people I interviewed at the summit, is the creation of effective preservatives to replace parabens. Curt Valva explained the problem succinctly: "Many of our products are water based. Things with high water content have to be preserved because as soon as you introduce water to something, bacterial growth starts immediately. You need something to either keep the bacteria from growing at its normal level, like we would have in drinking water, or kill it completely. The problem is those things that kill bacteria are also really not good for the human body. They're designed to kill cells—that's what they do." It's particularly important, Valva told me, to keep products like mascara, which are applied near the eye, clean and totally free from mould, fungus and bacteria.

Such is the interest in non-toxic preservatives that Judi Beerling led a summit workshop on Saturday morning dedicated entirely to this topic. It's not just the question of whether new technologies are available, but also whether they are cost effective. "You're often looking at double the cost. Sometimes triple. If you've got really high buying power, say you're a large multinational, you'd likely be able to get that down. It also depends on what you're trying to make: Some things are easier than others."

Though they were better looking and rather better dressed, the people in Beerling's busy Saturday morning workshop reminded me of those witches from *Macbeth*. All the participants formulated cosmetics as a profession, and as they traded knowledge about their favourite obscure plant ingredients, the witches' "Eye of newt, and toe of frog, wool of bat, and tongue of dog" incantation kept coming to my mind. But, of course, the information exchange at the workshop was designed to result in good and not harm. Even better, the benevolent discussion was fascinating. I found out that rather than parabens, natural and organic products have traditionally contained natural preservatives like grapefruit seed extract, though new materials and technologies are gaining acceptance. Beerling spent considerable time at her workshop outlining these so-called hurdle technologies, which involve the intelligent combination of different preservation factors to create a hostile environment for bacteria right in the product itself. The goal is to block growth of microorganisms by putting successive impediments in their path—each diminishing the population until none remain. Some of these approaches include using materials that make formulations ever so slightly more acidic or adding emollients with properties that can disrupt the membranes of bacterial cells. Other natural cosmetics are boosting their preservative systems through the use of antioxidants or by adding tiny amounts of spice extracts or alcohol. It turns out that one of the best—and simplest—ways to reduce the need for high levels of preservatives is to improve packaging. Smaller packaging helps to reduce or remove contamination issues. Airless dispensers or pumps can dramatically cut bacterial growth—unlike the huge, goopy Oil of Olay wide-neck skin-cream jar that sat on my grandmother's bathroom counter throughout my childhood.

Not that long ago, the cosmetic industry's knee-jerk solution to the problems of preserving the lipstick, shampoo and shaving gel in your bathroom was always the same: "Add more parabens! more parabens!" Now, through the leadership of innovative chemists like Judi Beerling, a larger number of less toxic options is now available in the manufacturer's toolkit. Just how many options became obvious to me when, as a way of concluding her Saturday workshop, Judi started flashing up ingredient labels from real-life natural cosmetics and invited the crowd to start playing the game "Spot That Preservative" by yelling out the answer. As I left the workshop to catch my plane, the cries of "honeysuckle extract," "tocopherol" and "thyme oil" followed me down the hall.

Beerling regards the recent success of sustainable cosmetics as something of a personal vindication. "I remember in 1978, my first business trip to the U.S. as a young chemist, I was put in front of a very senior VP at a multinational. He put me on the spot and said, 'So what's the next big trend then?'" She chuckled. "And I said, well I think natural is where things are going to go. And he said, 'Oh we've done that, haven't we? We've had all the green apple shampoos.' Ha! Who was right then?"

And Now, the Experiment

After doing a ton of research into the alleged merits of green personal-care products, I decided that some more direct testing of our own was in order. The idea was pretty simple. I asked for volunteers: Ray Civello of Aveda and Jessa Blades, one of *Glamour* magazine's 70 eco-heroes and the TreeHugger website's Best Green Makeup Artist for 2011, were up to the challenge. We wanted to look at the day-to-day differences in our participants' phthalate and paraben levels as they made the switch from using conventional chemical-laden, personal-care products to products that claimed to be greener and, notionally, safer.[25] The hitch with Ray and Jessa is that they'd long ago made the switch to natural products: They were already big believers in the notion that the first step in detoxing is to *avoid* harmful toxins. For the purpose of our experiment, we asked them to go back—just for a day.

Based on consultation with experts like Dr. Shanna Swan, we designed our protocol as follows.

On Day One, our participants had to undergo a 24-hour "washout" phase, which really just meant avoiding the use of any cosmetics or personal-care products as much as possible. The logic behind this washout is that chemicals like phthalates and parabens are excreted in the urine, usually within 6 to 12 hours of application/ingestion/inhalation.

After 24 hours of cosmetics-free living, our participants gave their first urine samples on Day Two at 8 a.m. That sample was used to establish their body baseline levels for the phthalates and parabens we were examining. Immediately following the first urine collection, Civello and Blades each did a one-time application of the conventional products we had sent them.[26] With the help of a study on chemicals in consumer products from the Silent Spring Institute[27] and the EWG's Skin Deep Cosmetics database,[28] we selected products that we were pretty sure contained phthalates and parabens aplenty.

Another urine collection was done at both the four-hour (noon) and the six-hour (2 p.m.) post-application marks, for a total of three urine samples from each for the first phase of the experiment. Following the 2 p.m. collection, we again asked our volunteers to refrain from using any more cosmetics for the duration of the second washout, which would end at 8 a.m. on Day Four. The second phase of the experiment played out the same as the first, with urine collections at 8 a.m., noon and 2 p.m. The only difference was that in this phase, Ray and Jessa would be applying the natural products listed in the notes for this chapter after their 8 a.m. sample.

So there you have it. After a total of 8 days, 12 samples and over 50 cosmetic products (oh, and a somewhat sleepless night as I worried about the impact Hurricane Sandy would have on New York City and subsequently the fate of Blades's urine in her Brooklyn freezer), the urine was packed up and sent off to a lab in British Columbia to be analyzed for various phthalates and parabens.[29]

Our participants were good sports about the whole thing. When I spoke to Civello as he was doing the first washout, he told me he was interested to see if his colleagues would notice: "I've smelled the same way for 25 years. I have a distinctive aroma that people know me by. People usually say to me, 'Man, you smell good!'" The day he walked down his office hallway stinking of Axe body spray rather than Aveda's distinctive rosemary mint, he did indeed turn some heads.

"We're really confused as to what clean smells like," Blades quipped over the phone, having a hard time adjusting to her new-found synthetic scent. Blades doesn't normally wear strong perfumes, as she's wary of their undisclosed ingredients. But as a professional makeup artist, she concedes that there's no question that long-wear conventional products work. "If you put plastic into a product, it'll stay on. Put some of this lipstick on the back of my hand and it stays there. But women don't have to wear waterproof mascara every day. They have to wear mascara that doesn't burn their eyes and that doesn't make their eyelashes fall out."

The results of our experiment were very convincing (check out Figure 4 for the story in a nutshell).

Figure 4. Graphs of Jessa's and Ray's MEP and methyl paraben levels (in ng/mL) over the course of the four-day experiment. Note the peak in the levels of both chemicals after application of conventional products.

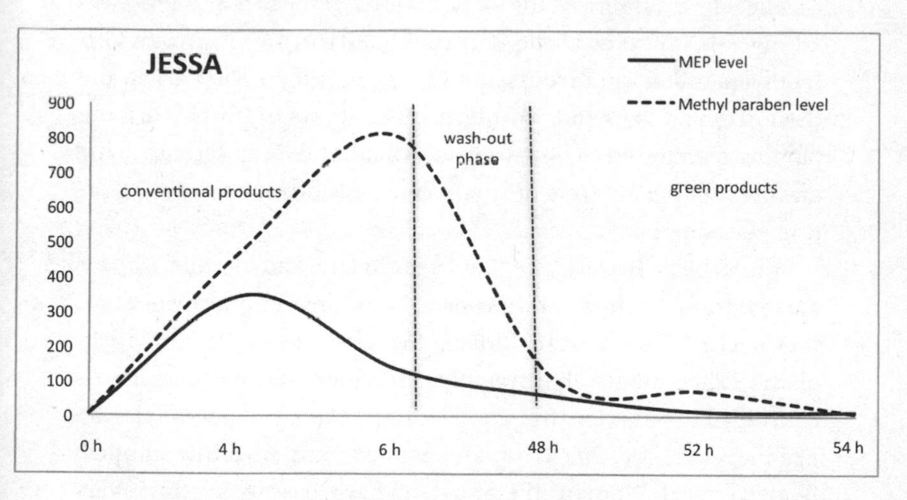

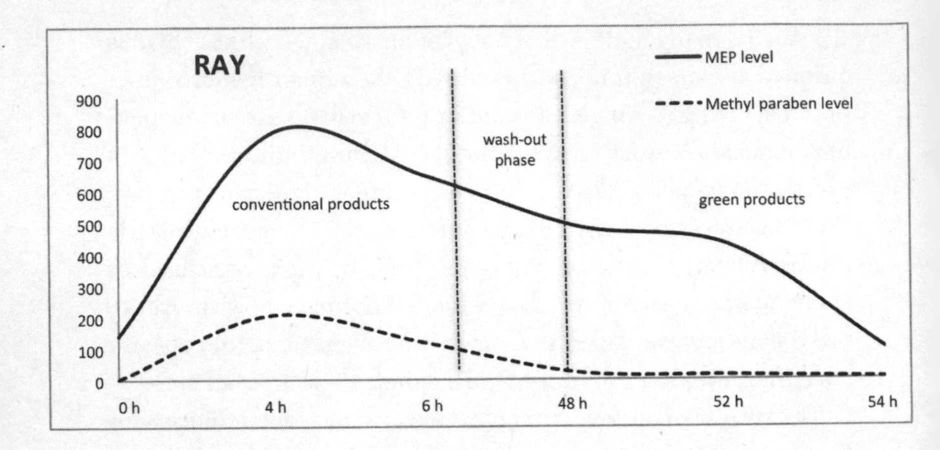

Blades's initial level of mono-ethyl phthalate (MEP) went from a low of 6.09 ng/mL to a high of 346 ng/mL then back down to 12.6 ng/mL. Her methyl paraben levels started at 5.17ng/mL, went all the way up to 805 ng/mL and then dropped to 7.69 ng/mL. Civello's levels followed the same pattern: His base MEP level was 143 ng/mL, it peaked at 786 ng/mL and fell to 99.3 ng/mL, and his methyl paraben went from a low of 2.45 ng/mL to 206 ng/mL and then dropped back to 4.29 ng/mL. The levels of phthalates and parabens we found in our study match other data in scientific literature looking at levels of these chemicals after topical application of cosmetics.[30]

While both Jessa's and Ray's phthalate and paraben levels spiked dramatically after their use of the conventional cosmetics, they declined dramatically during the subsequent washout phase of the experiment. Significantly, phthalate and paraben levels continued to fall even after application of the green personal-care products. The results from this experiment certainly support Blades's point. Women and men don't have to wear this stuff every day. Nor do they have to wear sackcloth and go without cosmetics entirely to avoid synthetic chemicals. The green products used by Blades and Civello are wonderfully effective. And they don't leave a toxic residue in the body.

Harvest Time

Here's a bonus: Not only are organic and natural personal-care products better for you personally (as our experiment convincingly demonstrates); it also turns out that they're better for the earth.

In the "good old days," when all cosmetics companies made their products with petrochemical derivatives, you could just phone up your ingredient suppliers and get them to cook you up some new toxic goo in their lab. Easy. Predictable. And fast. With the rise of natural cosmetics, however, an increasing number of companies are on the hunt for previously obscure plant-based ingredients. And a new and fascinating symbiotic relationship has emerged between the producers of these plants and the companies they serve.

I was first tipped off to this phenomenon during the keynote that Horst Rechelbacher delivered at the Sustainable Cosmetics Summit. At one point he made a crack about jojoba oil. "And those of you who use jojoba oil know how hard that's been the last few years." I looked around, puzzled, as knowing looks and rueful chuckles swept the crowd. I later found out that prices of jojoba oil had been volatile of late, in one year doubling or more, causing much angst for companies like the ones assembled. The intensifying scramble to source reliable supplies of plant-based ingredients emerged as a preoccupation with a number of people I spoke with.

Curt Valva of Aubrey Organics told me that there had recently been a shortage of blue chamomile oil. "It comes from Morocco," he said. "The crop was just devastated, we got very little and we use a lot of that particular ingredient. Our price doubled. And when a raw ingredient that is a mainstay of your product line shoots up in price like that, it becomes very, very difficult, let me tell you!"

Weleda's Jasper van Brakel was worried about organic pomegranate seed oil. "We use a lot of pomegranate seed oil from Turkey. We press the seed and use that in face-care products. Apparently, it's the best of the best—antioxidants, vitamins—so if that harvest fails because there's a drought or something, we're screwed."

Many natural cosmetics companies prominently mention their partnerships with traditional communities as tangible indicators of their authenticity and commitment to fair trade and sustainability. Aveda has created a whole sub-brand, "Soil to Bottle," around its tracing of raw ingredients back to their farm or harvesting co-op of origin. According to the Aveda website, the company's purchasing has resulted in improved standards of living for subsistence Indian farmers (turmeric), Australian Aborigines (wild sandalwood oil) and the Yawanawa people of Brazil (urukum seeds).

One of the ingredients most important to Bill Whyte at Badger is organic olive oil. They use hundreds of litres of the stuff every year, which they buy from a single little company—Soler Romero—in Andalusia's Jaén province in southern Spain. I spent some time chatting on the phone with Mónica Marín, the Export Manager for Soler Romero, from her office in the middle of the farm's six-hundred-hectare organic orchard. With her hundred-year-old trees crowding outside her window, Marín brought me up to speed on the global olive oil industry.

"I bet you thought that most olive oil comes from Italy, right?" she asked, the frustrated tone in her voice making clear that "Italy" was most definitely the wrong answer. "Not true! Spain is the biggest producer of olive oil in the world. We produce three times what Italy does. But they've got the brand!" The reason that I couldn't ever recall seeing "Extra Virgin Spanish Olive Oil" in the store became clear after Marín explained that Italy actually imported massive quantities of olive oil from other countries, mixed them together, then resold it as Italian product. Until recently, this had been the fate of the majority of Soler Romero's crop. In the same family since the 1850s, Soler Romero had always sold its olive oil in bulk to the local co-op. "Then 10 years ago," Marín explained, "the family decided to change the business. We received organic certification and built our own olive mill." The newly rebranded organic product soon came to the attention of Bill Whyte back in New Hampshire.

As Badger's business has grown, so have its orders from Soler Romero. Marín estimated that Badger's orders have increased by 40 percent over the past few years. Though the bulk of her oil is still sold for human consumption, the component taken by Badger for cosmetics is increasingly important. From the point of view of Soler Romero's owners (the seventh generation of their family to try to make a go of it in the highly competitive international olive oil business), every litre of product they are able to sell to Badger is a litre they don't have to offload, at substantially lower prices, to the bulk Italian market at the end of the season.

Another businesswoman who is grateful for the growth of the natural cosmetics market is Eugenia Akuete, President of the Global Shea Alliance. On the phone from Accra, Ghana, where she owns a shea butter–producing company, Akuete is bullish on the growth of her industry. "Shea nuts are harvested from across 21 sub-Saharan African countries," she told me. "We're a big, a growing industry. Every year, an estimated 1 million tons of shea nuts are harvested by 16 million women." This latter figure was so astonishing, I asked Akuete to repeat it. She confirmed that, yes, I'd heard her right. Sixteen million African women depend on income from harvesting shea nuts every year. The Global Shea Alliance was created in 2010 to bring together various stakeholders to build capacity in the industry.

Shea nuts are one of those perfect natural creations that have the power to make even the most hardened non-believer consider the possibility of a divine plan. If you boil them up, you can make a highly nutritious butter that has been an important food source for centuries. Rubbed on the skin, the vitamin E–rich oil is a useful natural moisturizer. Best of all, the nuts are produced by slow-growing, hardy trees throughout the arid plains of mid-Africa in an area where agriculture is tough slogging at the best of times. The size of a large plum, the fruit ripens at an opportune season of the year when other food supplies are at their lowest ebb. The thin pulp is tart but edible, and the large nut inside contains a veritable cornucopia of nutritious oils and fats. Much of the annual harvest is consumed by millions of households locally, but it's now also a valuable export commodity: as an ingredient in foods such as chocolate (where it is often used as a substitute for cocoa butter) and, increasingly, as a base for natural cosmetics.

In 1994, 50,000 tons of shea nuts were exported from West Africa. By 2012, that number had increased fivefold. From a standing start in the mid-1990s, about 24,000 tons of shea nuts are exported for cosmetics each year, a number that continues to increase rapidly.[31] I asked Eugenia Akuete whether, given that it's a semiwild crop and supplies are presumably limited, there is further room to grow the shea harvest. "Yes!" she replied. "Every year, an estimated 1 million tons of shea nuts are harvested. And we think there's an additional 1 million tons that are not used and left to rot."

Akuete told me a funny story about trying to get her company's shipment of shea butter through the American border in the 1990s. "Nobody knew what the product was," she said. The process took a while. More recently, at the same airport, Akuete was trying to get a shipment delivered, and before she could answer any questions, another border agent overheard the conversation. "You don't know what shea is?!" they asked incredulously, before taking it upon themselves to educate their colleague about the miracle butter's many virtues.

With fans like that, it's clear this grassroots industry is only getting started.

The New Normal

When I started at the University of Guelph in 1987, I arrived on campus determined to delve deeply into academic lefty culture. Guelph, still one of my favourite towns, made this easy. "The Royal City," as it's known, has long been an island of latter-day hippies adrift in suburban southern Ontario. Parties at the time would often end with "Rise Up," the gay rights anthem by Toronto's Parachute Club, or Midnight Oil's "Beds Are Burning." We rocked out at U2's

Joshua Tree tour with socially conscious songs like "Bullet the Blue Sky." I got arrested at peace and environmental demonstrations, grew my hair and sideburns really long and determined that I should buy more organic food and products. At the time, the only place in town to purchase the requisite organic granola was at the food co-op. This was a dank, subterranean room—think Rome, Christians and Catacombs—for which you needed to pick up a special key at the business upstairs. Co-op members were required to volunteer to staff the place one night a month: Work the cash register for other members doing their shopping, sweep up and clean up the oozy dribbles on the floor from the bulk soap and laundry detergent dispensers. It was a lonely gig. I can't remember ever meeting another living soul during my shifts.

My point in taking you on this stroll down my memory lane is twofold. First, in the space of a short (I mean, 1987 wasn't that long ago, right?) period of time, all things organic and natural—cosmetics, food, cleaning products—have become much more available. All my neighbours in my downtown Toronto neighbourhood and I can now get many of the goods we need from the substantial organic and natural section of our local Loblaws—one of the 1,400 supermarkets run by Canada's largest food retailer.

Secondly, though it wasn't the Summer of Love by any means, back in 1987 there was still a tight correlation between how you looked (long hair and tie dye), how you cooked (did anybody really like the brown rice in that damn *Enchanted Broccoli Forest?*), what you thought (Nicaragua's Sandinistas were the cat's meow) and even how you smelled (patchouli still turns me on). Today? Not so much.

"The appeal of green cosmetics is super broad," says Adria Vasil. "I get questions from yummy mummies made up after yoga, from businesswomen, from guys in suits—it's definitely changed. I think we've still got a little further to go to penetrate the market when it comes to the beauty bunnies who are using the most cosmetics who really don't think the options are out there, but I'm here to tell them that yeah, those products are out there. If you want that vixen red lip and that smoky eye, no problem."

Is anything standing in the way of increased growth of sustainable cosmetics? No. In fact, it's quickly becoming the new normal. Within a few short years, most personal-care products will look a lot more natural. Are people changing everything overnight? Of course not. Florida-based pollster Linda Gilbert's research has shown that people usually change up their brands in a gradual way. "In our discussions with consumers, we find that perhaps counterintuitively, conventional products and green products will co-exist in the bathroom cabinet," she told me. "It's the same phenomenon that we observed 20 years ago when we realized that just because someone buys organic cereal doesn't mean they'll buy organic milk. People really do pick and choose, and it's always this balancing act between brand trust, performance and considerations related to health and environment. So they may use Tom's of Maine toothpaste but also gargle with Listerine."

Bottom line? Gilbert has identified "three things that consumers tell us are getting in the way of buying more organic and natural products." The first is availability. The second is price. And the third is effectiveness or practicality.

As I hope has been obvious from reading this chapter, all three of these obstacles are well on the way to being removed. You no longer need to be an Old Order Mennonite to avoid phthalates and parabens.

TWO: ORGANIC TEA PARTY

~ Rick eats broccoli ~

Life expectancy would grow by leaps and bounds if green vegetables smelled as good as bacon.

—DOUG LARSON

"CAN I ANSWER 'ALL OF THE ABOVE'?" the young mother wearing yoga pants and a stylish purple barrette asked, her little daughter peeking out from behind her legs.

"Naw. That won't really tell me much," I replied. "Even if you agree with all these statements, which is most important to you?"

She looked again at the multiple choice question displayed on my clipboard as we talked in the middle of the canned food aisle of east end Toronto's organic food mecca: The Big Carrot. With an uninterrupted river of busy shoppers flowing past, we occasionally had to stand on our tiptoes to let them by. Having once been called "Canada's tallest environmentalist," I awkwardly held the clipboard over everybody's heads.

"I guess it would have to be number one then. I'm really concerned about toxic chemicals."

"Really appreciate your time. Thanks!" I turned to quiz the athletic-looking guy in the hoodie coming toward me, his arms laden with fresh produce, his mouth full of half a blueberry muffin, his face wearing a guarded expression of the type usually reserved for unwanted door-to-door solicitors.

It was just another Saturday afternoon at the busy grocery store in the middle of Danforth Avenue's Greektown neighbourhood. I had come to the epicentre of sandal-wearing culture in my city, survey in hand, with one simple question: Why do you buy organic food? Over the course of a few hours, my co-worker Rachel Potter and I spoke with almost 250 people, and the results were clear.[1] More than 60 percent of those we quizzed identified their concern with toxins in non-organic food as their main reason for shopping at The Big Carrot—a finding similar to those of other studies that have looked at this question.[2] Yes, some people gave other reasons (organic food tastes better and is higher in key nutrients; it's better for farmers and the environment), but overwhelmingly, more and more shoppers are "going organic" in an attempt to shield themselves and their families from toxic chemicals.

The Big Carrot itself is a testament to the increasing interest in organic food. Founded in 1983 when five workers in a natural food store found themselves out of work, "The Carrot," as it's affectionately known, was the first one-stop-shop natural and organic food store in Toronto. Though offering a full range of products, it survived for a number of years in a squishy two-thousand-square-foot space: a far cry from its current palatial digs in the sprawling "Carrot Common" complex, which includes sundry retail outlets like a juice bar, wholistic dispensary and independent bookstore. In its earliest days, the entire operation grossed about $11,000 in revenue over the course of a week. Now that's a bit less than the amount The Carrot makes in a single day on fresh produce alone. On average, the store services more than three thousand people every day and sales continue to increase at about 5 percent per year.

Comparing Apples to (Organic) Apples

So let's drill down to the main concern of Big Carrot shoppers: Sure, people want to avoid toxins. But is there any proof that organic food can deliver on this?

I'll put my cards on the table. Though I've long been a consumer of organic food, I'd always assumed that empirical evidence of its worth was lacking. In that case why did I buy it, you ask? Out of faith, I guess. A sort of nutritional Hail Mary Pass, a hope that it was sort of, kind of, better—if not for me, then for the planet.

Not everyone is so casual about this, however. Judging by the vehemence in his voice coming through the phone line, Chuck Benbrook has little time for such vague thinking. Organic food has hard science on its side, he tells me.

A professor in Washington State University's Center for Sustaining Agriculture and Natural Resources (CSANR), Benbrook has had a long career in the agricultural sector. For nearly 20 years, he worked in a variety of senior roles in the U.S. federal government and has published widely on pesticide use and residue levels in food, as well as on the relative nutritional merits of different types of produce. After a recent article that was played by the media as torpedoing the value of organic food (discussed further below), Benbrook dryly rebutted this assumption, noting that he was among a small group of people who had actually *read* a couple hundred of the food studies the article cited.[3] No two ways about it: Though his preferred reading material may lack diversity, the guy has done a lot of thinking about organic food, and he is unequivocal in his assessment. "There is absolutely no way to escape the conclusion, based on a mountain of data, that consumers who seek out organic food to reduce their own personal or family's exposure to possibly risky pesticides, are getting a significant return on their investment. They're lowering their risk by approximately 80-fold."

Benbrook has focused on pesticides in food because of the evidence that much of a typical person's pesticide exposure comes from the residue of agricultural chemicals on the fresh fruits and vegetables that we eat every day. He is particularly concerned that certain highly vulnerable segments of the population—including children and pregnant women (especially those in the three months before conception and in the first trimester of pregnancy)—are being exposed on a daily basis to pesticide levels that have the potential to trigger developmental problems. His argument rests not only on the presence of pesticides in certain types of produce, but also on their toxicity.[4] Benbrook's research team has done what he characterizes as an "extremely excessive" analysis of the U.S. Department of Agriculture's pesticide programme database, which encompasses 15,000 to 25,000 food samples a year, including an increasing number from organic food. They have analyzed these figures back to 1993 and have created the "Dietary Risk Index" (DRI), which takes into account the frequency of residues, average residue levels and toxicity of the pesticide in question.

For example, Benbrook and his team compared both conventionally and organically grown varieties of 11 tree-fruit crops and found that there was, on average, an 84-fold difference. That is, the pesticide residues that typically appear in conventional tree-fruit crops in the U.S. food supply make that produce 84 times riskier (to cause harm as defined by the United States Environmental Protection Agency) to consume than organic fruit. Benbrook has done the same sort of head-to-head comparison for many vegetables and concludes that the DRI difference is typically between 50-fold and 150-fold.

To boil this down further, Benbrook has translated the DRI into the typical experience of an individual consumer. According to his calculations, on an average day, an American consumer of conventionally grown produce will be exposed to 17 pesticide residues: a combined DRI value of 2.0. Now, if this consumer replaces 12 foods or ingredients in their diet with organic items, the number of residues will drop from 17 to 5, and their combined DRI value will drop by over two-thirds, to 0.62. This finding fits well with one of the key points of this book—that is, the first thing to do in order to get toxins out of your body is to avoid them in the first place. Before engaging in some of the detox therapies Bruce outlines in our later chapters, this is the place to start. If you don't have to ingest toxins, such as those found in conventional food, don't. You can replace your toxin-ridden produce with organic goodness.

One interesting phenomenon that has Benbrook increasingly concerned is the change in the relative contribution to consumers' pesticide risk of domestically grown versus imported produce. "In 1996, when the Food Quality Protection Act (FQPA) passed in the United States," he noted in an interview with me, "by our calculations about 75 percent of the dietary risk from pesticides came from domestically grown fruits and vegetables and about 25 percent of the risk came from imported foods. Today, the ratio has flipped to 75 percent from imports and only 25 percent from domestically grown foods." His conclusion is that though the U.S. Environmental Protection Agency (EPA) is using the 1996 law to reduce pesticide levels in U.S.-grown food—undeniably good news—this has no effect on how pesticides are used overseas. As a result, most of the exposure of American (and, we can assume, Canadian) consumers to pesticides like malathion and chlorpyrifos comes during the winter months when we rely on imported fresh produce. "If this trend continues, another five years down the road, we might have 90 percent of the risk to consumers from pesticides happening in three months, virtually all from imports," Benbrook concluded grimly.

So according to Benbrook's evidence, the answer to my initial question is clear: Most North American consumers are being exposed, on a daily basis, to pesticide levels that can hurt them. And unfortunately, others share Benbrook's view. More scientific evidence of this problem is accumulating by the day (see Table 5). Exposure to many commonly used agricultural pesticides is linked to a variety of health problems, particularly in kids.

It turns out that the Big Carrot shoppers have reason for concern after all.

Table 5. Recent science points to negative health effects from pesticide exposure

Health Effects	Study Description	Year of Study
Non-Hodgkin's lymphoma	Elevated levels of organochlorine (OC) pesticides years before cancer develops increase risk of non-Hodgkin's lymphoma later in life.	2012[42]
Endocrine disruption	Thirty out of 37 widely used agricultural pesticides blocked or mimicked testosterone and other androgens.	2011[43]
General developmental problems and cognitive deficits in infants and children	Prenatal exposure to chlorpyrifos associated with decreased IQ and working memory scores Increased levels of prenatal dialkylphosphates (DAPs) associated with lower neurodevelopment scores at 12 months Increased urinary dimethyl organophosphate (OP) pesticides associated with increased odds of ADHD	2010, 2011[44]
Low birth weight, smaller brains	Higher levels of OP pesticides in umbilical cord plasma associated with lower birth weight, shorter length and smaller head circumference	2011[45]
Reproductive problems	Exposure to fungicides—especially when in combination with other chemicals found in consumer products, food and the environment—is associated with changes in anogenital distance, sex organ weights and genital malformations	2009[46]

Health Effects	Study Description	Year of Study
Asthma and poor respiratory health	Any exposure to pesticides was associated with an increased risk of asthma, while living in a region heavily treated with pesticides presented the highest risk.	2011[47]
Increased risk of obesity and diabetes	Low-dose pesticide exposure predicts incidents of Type 2 diabetes and future adiposity, dyslipidemia and insulin resistance	2010, 2011[48]
Infertility	Increased concentration of DDT in maternal milk was associated with an increased risk of infertility	2009[49]

The Birds and the Bees and the Kids

If pesticide residues in organic produce are significantly lower than in conventional produce, the next question to answer (given the vagaries and variability of the human metabolism) is whether the transition to organic food will result in measurably lower pesticide levels in the body. Like Ray and Jessa's experiment with cosmetics in Chapter 1, does the notion of "detox by avoidance" work for pesticides in food as well?

Of the surprisingly few studies that have explored this area, the most convincing is that of Alex Lu, a professor in the Harvard School of Public Health in Boston.[5] I sat with Dr. Lu in his sunny office in the beautifully restored art deco Landmark Center as he gave me a short history of pesticides. "Every 30 years or so we start using a different kind of pesticide," he said. "There are a couple of reasons for this: One is because of resistance—the more you use a pesticide, the more tolerant of it the insects become. The other reason is that negative effects on human health or the environment start to surface." Qualitatively speaking, Lu continued, our grandparents were exposed to different pesticides than we are today. Theirs was the generation that got bombarded by organochlorines—the family of compounds that includes DDT. Many organochlorines have subsequently been banned around the world because of the nasty effects documented in *Silent Spring* and elsewhere: animal die-offs, links to human disease and extreme persistence in the environment.[6] "After the banning of DDT we started using organophosphate pesticides," Lu said. "And around 2000, the U.S. started taking action to reduce organophosphates because evidence of health effects began to surface. History repeating itself."

Though the U.S. Environmental Protection Agency had dramatically reduced most indoor uses of organophosphate pesticides by the early 2000s, the use of chemicals like malathion and chlorpyrifos on food crops was still allowed. Lu decided to investigate the effect of this practice on the human body with a simple but elegant experiment. "At the time nobody knew whether eating organic produce would have any beneficial effects, specifically for pesticide exposures. So we designed a crossover study with a group of kids lasting 15 days. For the middle five days, they ate nothing but organic produce, and they ate their regular conventional food before and after the organic feeding days. What we wanted to find out was whether the eating of organic produce would reduce the pesticide exposure that we measured in the urine samples." After completion of the study, Lu and his colleagues found that not only did the organic food they provided reduce the kids' pesticide exposure, it largely eliminated the residues of malathion and chlorpyrifos in their bodies. The study demonstrated that an organic diet provides almost immediate protection against exposure to organophosphate pesticides that are commonly used in agricultural production. And it also showed that children are most likely exposed to these particular chemicals exclusively through their diet.

Interestingly, when Lu tested the same group of kids for another class of pesticides—pyrethroids—he found no similar reduction in levels as a result of eating organic food.[7] Pyrethroids are synthetic chemicals similar to those found in chrysanthemums. Because of their effectiveness in eating away at the hard external skeleton of various bugs, they are now the most commonly used pesticides in the average home. Though the levels of pyrethroid pesticides in the urine of Lu's test subjects didn't diminish as a result of eating organic food, kids whose parents reported using pesticides at home did have significantly higher pyrethroid levels. From this Lu concluded that residential pesticide use (both in the house and in the garden) represents the most important risk factor for children's exposure to pyrethroid insecticides.

Alex Lu's latest work relates to an even newer family of pesticides that have started to replace some organophosphates: neonicotinoids.[8] Synthetic chemicals similar to nicotine, neonicotinoids are "systemic insecticides": Rather than sitting on the surface of the crops, they are actually slurped up by a plant through its root system. As Lu explained, right now Monsanto is teaming up with Bayer AG to coat their GMO corn seed with neonicotinoids. The claim is that, for the farmer, it will be one-stop shopping: You buy GMO corn pre-treated with insecticide and you don't even have spray. You just put down the seed and it will grow. The downside is that in the past when farmers sprayed with pesticides, consumers hoped that the residue on, for example, the surface of a tomato would dissipate over time because of sunlight, rainfall and so on. But with this new technology, Lu pointed out, chemicals actually reside *inside* the tomato. There's no way that you can remove the chemical prior to eating it.

Luckily for tomato eaters (and eaters of any other conventionally farmed fruits and veggies), there is currently no evidence that neonicotinoids are hazardous to human health. A good thing, given that one chemical from this family, imidacloprid, is the most commonly used pesticide in the world today.[9] It's not so lucky for one group of insects, though: the bees. There is now considerable evidence that imidacloprid causes "colony collapse disorder," or CCD—the mysterious, global trend whereby bees up and desert their hives, never to be seen again. Given that bee pollination is involved in foods that comprise one-third of the U.S. diet and that commercial pollination is valued at about US$217 billion a year globally, this is a bit of a problem, to say the least.[10] There are many competing hypotheses being put forward to account for CCD, which has now manifested itself in many places around the world. Some have theorized that colony collapse disorder may result from mite and virus infestations, bee malnutrition and even cellphone radiation.[11] In his latest work, however, Alex Lu provides convincing evidence that very small doses of imidacloprid can produce the CCD effect. In his carefully constructed experiment, 15 of 16 imidacloprid-treated hives were abandoned in 4 different apiaries 23 weeks after imidacloprid dosing. The survival of the "control" hives—those not treated with insecticide and managed alongside the treated hives—buttresses this conclusion.

In a serious bummer for them, honeybees don't have the option of relying on organic food sources. They're stuck buzzing around crops which, it is now fairly clear, are slowly poisoning them to death. Luckily for them, as we were finishing up this book, governments started to step in to protect the bees (or at least their economic value). In late April of 2013, the European Commission passed a proposal to restrict the use of three types of neonicotinoids (including imidacloprid) effective December 2013 for a period of two years. The commission took action in response to what they identified as "high acute risks" for bees when exposed to these pesticides.[12] The United States Department of Agriculture initially rejected the EU claim of pesticides primarily causing colony collapse disorder, but called for a further investigation into the problem.[13] And the EPA is currently reviewing some neonicotinoids for re-registration. In Canada, the federal government is aware of and evaluating the issue, but better action is happening at the provincial level. An expert review panel was convened by the Government of Ontario to investigate neonicotinoids as the cause of honeybee deaths, and to make recommendations on the future use of the pesticides.

The Hundred-Year Diet

Firestorm. That's the word I'd use to describe the reaction to what became known as the "Stanford study" (reflecting the study's institutional origin) that hit the media as I was writing this chapter.[14] The title of the thing—"Are Organic Foods Safer or Healthier than Conventional Alternatives?"—tells you everything you need to know about the question it posed. And the screaming headlines that ensued—"Organic Food Isn't Healthier and No Safer," "Little Evidence of Health Benefits from Organic Food" and "It's Official: Organic Food Is a Waste of Cash"—were the organic industry's worst nightmare.

When I asked him for his reaction to the furor, Matt Holmes, the Executive Director of the Canada Organic Trade Association (COTA), let out an exasperated sigh and fiddled with his stylishly skinny tie. Holmes, the dapper, 30-something main lobbyist for the Canadian organics industry, literally lives and breathes the ups and downs of the organic sector. His wife, Beth McMahon, for many years ran the Canadian Organic Growers—the representative of organic farmers—meaning that a significant chunk of the Canadian organic industry's advocacy efforts has been conducted out of their living room in Sackville, New Brunswick. Holmes is generally very optimistic about organic's prospects. After all, since he started at COTA, Canada has implemented the first-ever national standard and labelling system for organic products and has signed major agreements with the EU and the U.S. to increase exports of Canadian organic food. Still, Holmes admitted to the occasional frustration. "I don't think there's an evil conspiracy against organic," Holmes told me as we shared a drink overlooking Baltimore harbour's historic warships. (We were both in town to attend Natural Products Expo East—one of the biggest organic industry conventions and trade shows of the year.) "I just think there's a human tendency to assume organic should be a solution to everything, and it's not. It can't be. But the media, in particular, love to pull the 'gotcha' routine."

Holmes told me about a recent incident involving a TV story about pesticide residues on organic apples. No matter how many times Holmes tried to explain that pesticides are ubiquitous (and therefore it's no surprise that even organic apples have a bit on their skin), the journalist was determined to run with an "exposé" angle: "Revealed: Pesticides Found in Organic Produce!" "The levels they found were similar to those in unborn children and Arctic sea ice," Holmes said. "The reality of today's world is that this stuff is absolutely everywhere. Organic doesn't promise to be completely pesticide free—it can't escape the random pesticides that are floating around. It promises to be an alternative system that doesn't contribute to the problem."

Holmes pointed out that the Stanford study set up a straw man that was further torqued through the extensive media coverage: that organic food is primarily desirable because of the claim that it has a higher nutrient content than that of conventional food. "It ignored or downplayed the other benefits of organic production. I like to describe organic agriculture as the hundred-year diet. It's a system of agriculture that perpetuates itself, that creates a healthy ecosystem that will in turn continue to support plants in the long term, so you're not in this deathly cycle of creating short-term nutrients—which then can contribute to pest infestations that need to be counteracted by immediate and short-term chemical pesticides, which then kills the life in the soil, which then requires another synthetic input. Just like we need to give our bodies the right tools and conditions to do their detoxifying jobs, organic tries to enable and facilitate the natural predators and the natural nutrition and micro-flora and fauna that should be in the system." Holmes underlined that, despite the media hype, buried in the Stanford study were tepid admissions of the worth of organic food: a finding that organic produce has a 30 percent lower risk of containing pesticides than does conventional produce (a number that Chuck Benbrook believes is closer to 90 percent[15]) and another finding that conventional meats have a 33 percent higher risk of contamination with bacteria resistant to antibiotics than do organic meats (a figure that Benbrook calculates as being far too low).

Even if you accept the Stanford study's contention that evidence for the nutritional superiority of organic food is lacking (which, by the way, flies in the face of other, more compelling studies that have concluded that organic foods are more nutrient rich[16]), it strongly reinforces the argument that organic food is better for the consumer when it comes to levels of pesticides and disease-causing bacteria. This, from a study that was widely trumpeted by the media as being a disaster for the organics industry.[17]

To quote the immortal words of Meatloaf's 1977 schlock-rock classic: "Two Out of Three Ain't Bad."

Post-Niche

Forget the Tea Party, that hard-nosed political insurgency that has, in the past few years, turned U.S. politics on its head. The Tea Party is yesterday's news.

The newest phenom on the block is the Organic Tea Party.

"In Organic We Trust" read the placards of the crowd milling outside the front doors of the Baltimore Convention Center. As I pushed through the picketers to get inside, one of them thrust a small envelope into my hand. It was a package of tea, a determined Uncle Sam staring out at me from one side and the Organic Tea Party's "platform" emblazoned on the other: "The Organic Tea Party is a nonpartisan celebration of people, planet and pure tea. Our aim is a healthy and happy planet where our food chain and tea farmers are free from pesticides, chemically-enhanced fertilizers and genetically modified organisms (GMOs) . . . Declare Your inTEApendence and Vote Organic!" In the 2012 U.S. presidential election season, this fun gimmick by the Numi tea company was the first of many indicators at Natural Products Expo East that the organic industry was feeling pretty pleased with itself. One might say even cocky.

During the conference in Charm City, I heard a keynote speech by Kathleen Merrigan, who was Deputy Secretary of the U.S.

Department of Agriculture at that time and—more than 10 years earlier—one of the lead authors of the USDA's organic labelling rules. In her upbeat remarks, Merrigan at one point said that the "USDA as a department is really owning organic," which caused the assembled organic industry leaders to look at each other and murmur in delight—no doubt thinking that only a few years earlier, such a powerful validation from a government leader would have been utterly unthinkable. Another, made-in-Canada surprise occurred at the gala dinner of the Organic Trade Association (OTA) (affiliated with Matt Holmes's COTA), the organic industry's lobbying arm. In a sea of (charming) Americans, I sat at a table with a number of fellow Canucks, including a guy from Export Development Canada (EDC). EDC is the Canadian federal government's export credit agency, and it's a big deal—providing financing to Canadian exporters and investors in more than two hundred markets worldwide. In 2009 alone, EDC facilitated over C$82 billion in investment, export and domestic support. Every few years EDC picks a new small batch of strategic areas on which to focus its largesse—areas that the agency believes will be winners on the international stage. And guess what? Organic agriculture is now one of those areas.

Another organic expert I chatted with in Baltimore (over a delicious cup of organic tea) was Katherine DiMatteo. The energetic and grey-haired DiMatteo was, for 16 years, the Executive Director of OTA, leading the industry through the critical years that culminated in the implementation of the first U.S. National Organic Program (NOP) standards in 2002. DiMatteo still remembers the excitement of that time. "It was a very unified effort. It was initiated by environmental and consumer activist groups, then the farmer organizations came on, and the Organic Trade Association—which was little bitty at the time—with our small group of processors and retailers also joined in. It was multi-stakeholder and multi-interest, but we knew this was *the* moment to get a breakthrough at the national level." The results of the new standard have been truly impressive. In 1990, the year that DiMatteo joined the "little bitty" OTA, U.S. sales of organic food were US$1 billion annually. In 2002, the year the NOP standards were implemented, that number had grown to US$8.6 billion. In 2011, U.S. sales stood at $31.4 billion.[18]

Despite this astounding growth (which shows no signs of abating anytime soon) and the generally positive mood in the Baltimore Convention Center, DiMatteo confessed to fretting about some storm clouds on the horizon. I asked her about a recent article in the *New York Times* that detailed infighting between organic advocates. The dispute has centred on the actions of the National Organic Standards Board (NOSB) in approving some food additives—like carrageenan, a seaweed-derived thickener—as being okay to include in certified organic products.[19] "Absolutely benign and consistent with organic principles," say the majority of those on the NOSB. "A sellout to big business and a betrayal," retort some organic purists like the Cornucopia Institute, publisher of the provocative report *The Organic Watergate—Connecting the Dots: Corporate Influence at the USDA's National Organic Program*.[20] Looking at me over the top of her glasses as she nursed her steaming tea, DiMatteo observed that the organic industry is often its own worst enemy. "Certified organic is a rigorous adherence to now nationally required standards," she explained. "It's costly in terms of certification, inspection, paperwork and all these other things, and the regulations constantly get tightened. Our movement wants continuous improvement, quickly—and we're not patient."

DiMatteo is concerned not only that this "constant tightening" of the NOP standards dissuades potential new entrants to the organic market, but also that the constant push to expand the meaning of "organic" is making the concept too unwieldy and undeliverable. "Other agendas are beginning to be played out in the organic regulations," she told me. "Like allergies. Our standards aren't about allergenicity: We're supposed to be about environmental protection and biodiversity. Now all of a sudden we have to talk about things like 'Is it an allergen?' The organizations that have the energy to push and work on these things are now using organic as a platform for moving towards an agenda that is making it very difficult for both the people already in organic to stay in and new people to get in." DiMatteo wonders whether some organic advocates are really trying to drive processed foods out of the organic certification system entirely. In her view—and I have to say I agree with her—this would be entirely counterproductive. "At least here in the U.S., we still buy an awful lot of processed food—that culture isn't going to change," she said. "The way to mainstream consumers is not by telling them you have to *only* eat fresh fruits and vegetables and whole grains."

It falls to DiMatteo's successors at the OTA to work through these complicated issues. Clearly, they have their plate full. The way Laura Batcha, the OTA's current Executive Vice President, sees it, the organic industry is at an exciting but vulnerable moment. "A lot of the conventional agriculture industry has been annoyed hearing the USDA endorse and recognize organic across the whole agency. Even though the institutions and the dollars and a lot of the change haven't happened yet, they just don't like the talk of it," she told me as we tried to find a quiet corner on the noisy convention floor. "There's been a much more coordinated pushback on us from the conventional industry, and we're at a vulnerable spot in our growth because we're no longer under the radar."

"So you're not niche anymore," I offered.

"We're not niche. We're post-niche! And it's a vulnerable place to be because we're between. We're increasingly mainstream, but we're still operating from a minority position."

Batcha suspects that, quite simply, the conventional agricultural industry has some serious control issues. "Over the last 70 years, agricultural policy has really been driven by commodities. If you produce it, you can sell it. And the production base and the technology behind production dominated what went in the market. Now we're at a place where food is driven by consumer choices."

This is a shift that's uncomfortable for folks who are operating out of a production mentality, she says. "Because bottom line: Organic is closer to the consumer than commodity agriculture is."

"Everywhere Food Is Sold"

Whether you're Apple or a small organic food store, the pressures and complications that emerge from expanding your business are often difficult to manage in any organization. Given the organic sector's very rapid emergence in the consumer mainstream, it's perhaps not surprising that there are some growing pains. Many organic companies are experiencing year-over-year double-digit increases in their revenue. Some of their stories are quite spectacular.

Here's an example. At its founding in Vancouver in 1985, Nature's Path sold just one product: Manna Bread, a dense and nutritious sprouted-grain concoction that quickly became a health food store fixture. Still privately held and run by Arran and Ratana Stephens and their children, Nature's Path is now the largest organic cereal company in North America, with annual sales in excess of C$200 million and distribution in 42 countries. As I munched my way through the overflowing basket of samples sent to me by Maria Emmer-Aanes, the company's Director of Marketing and Communications, I didn't find this terribly surprising. Their cereal is delicious: I now keep a pack of Love Crunch Apple Crumble granola on my desk. Emmer-Aanes told me that Nature's Path has bought nearly three thousand acres of farmland in Saskatchewan to supply many of their whole grain needs and continues to invest in social causes like the Prop 37 battle (described below). "The company is fiercely independent," she says. "We feel that it's socially responsible to make sure that everyone—no matter your economic status—has access to chemical-free food. This is a company that decides what its margins are going to be, decides what it's going to put back in the world and isn't afraid to make unconventional financial decisions. The strategy has paid off! When you give, it comes back to you."

Another story of success is that of possibly the most ebullient champion of all things organic: Gary Hirshberg, Chairman and former "CE-Yo" of Stonyfield, an organic yogourt maker in Londonderry, New Hampshire. A tagline on Stonyfield's website reads: *Environmentalists turned yogurt makers*—a succinct summary of the company's DNA. Hirshberg joined what was a tiny business in 1983 hoping to turn the cottage industry into a profit centre that would fund a local farming school dedicated to teaching sustainable agricultural practices. That year, the first 50-gallon batch of yogourt was made, and annual sales totalled US$56,000.[21] The demand for their yogourt soon outpaced Stonyfield's 19 cows' ability to supply the milk, and the company began buying product from local dairy farmers. By 1988, the company had expanded so much that they were able to build a large industrial facility in Londonderry, New Hampshire, with sophisticated dairy-processing equipment that could handle the increasing demand.[22] Today, Stonyfield Farm has production companies in the U.S. and Canada and ventures in France and Ireland—and is valued at US$400 million.[23] Choosing a different growth strategy than Nature's Path, Stonyfield has partnered with food product multinational Danone Group, which now owns 85 percent of Stonyfield shares. This international-scale transaction has helped propel the company to its current position as the largest organic yogourt producer in the world and the third-largest yogourt brand in the United States.[24]

Stonyfield's affiliation with a larger global food company is typical of a wider consolidation trend in the organic sector. Multinational food companies have got into the business of not *going*, but *getting* organic: acquiring successful, smaller organic companies. Those delicious, all-natural "Naked" fruit juices and smoothies? Since 2006 these beverages have been brought to you by PepsiCo. General Mills, the makers of such staples as Betty Crocker, Lucky Charms and Pillsbury, actually entered the organic business pretty early on, with their 1999 acquisition of Small Planet Foods, adding the likes of Muir Glen organic canned tomatoes, juices and sauces to their already humongous stock list. And what better example of a granola going corporate than the buyout of Kashi by the Kellogg Company in 2000?[25]

For Hirshberg, a wiry and intense guy who admits to preferring peppermint tea because no one could handle him on caffeine, there's no time to waste. "The transition to organic has to happen faster. Yes, our sector has to evolve, but we've got to become part of the mainstream. It's got to happen. We're on a chemical time bomb as a species." Hirshberg observes angrily that 50 years after Rachel Carson's publication of *Silent Spring*, companies are still using the same arguments to discredit her. "The legacy of this multidecade experiment with chemicals is what Al Gore likes to say: IBG YBG—I'll be gone; you'll be gone. That's what will happen if we keep following the credo 'Just make your money, get your margins and get out of Dodge with your stock options.'" Hirshberg's perspective is different and simple: "Everywhere food is sold there should be organic. Walmart, airlines, airports. I've worked very closely with Walmart—I think I was the first mainstream organic company to be in Walmart. I have gone down and given lectures at their infamous Saturday morning town meetings. I've met with heads of Walmart in other countries. I think a lot of activists are very troubled by places like Walmart, but I'll tell you the nail in the coffin of synthetic growth hormone was laid by Walmart." In Hirshberg's view it was Walmart's refusal to buy products containing synthetic growth hormone that spurred other companies, like Danone and Yoplait, to change. "A big player like that can move mountains."

Given that Bruce and I have the primary objective of reducing levels of synthetic chemicals in people's bodies—as many people as possible as quickly as possible—I tend to agree with Hirshberg's logic. After chatting with him, I actually phoned up Walmart to find out more about their sales of organic products. Just by dint of its enormous size (accounting for about 20 percent of all groceries sold in the United States), Walmart moves more organic food than any other retailer.[26] I spoke with Ron McCormick, Senior Director of

Sustainable Agriculture for Walmart U.S., at their Bentonville, Arkansas, Home Office. He told me that after a very public commitment in 2006 to double sales of organics, the retailer has now zeroed in on those areas where organic sales continue to grow.[27] "If you look at where organic is really selling, you have big dollars in dairy and big dollars in organic baby food and big dollars in fresh produce. In all the other departments, there are considerable dollars just because we have 4,500 stores around the country, but in any individual store you wouldn't see anything particularly significant."

Local vs. Carbon

I was surprised when McCormick told me that these days, Walmart sees more interest in local food than in organic. The retailer has made a commitment to increase local's share of its total produce sales to 9 percent by 2015—and it's on track to blow past this target much sooner. McCormick foresees further great opportunities to increase the amount of local production across the country, and in this sense Walmart is simply responding to what is clearly a pervasive trend. "Locavore" was named 2007 Word of the Year by the Oxford American Dictionary, and I now stumble upon farmers' markets selling local produce in virtually every city I visit. Locally sourced ingredients are now *de rigueur* on restaurant menus everywhere, and "100 Mile Diet" has now entered the vernacular.

This rise in the prominence of local food clearly irks some organic advocates. The Canada Organic Trade Association, for instance, has published a hard-hitting postcard enumerating the advantages of organic food: It's subject to a rigorous, government-regulated, labelling and certification system; it's guaranteed to have been produced without the addition of pesticides, synthetic fertilizers or growth hormones or antibiotics; it's made without artificial preservatives, colours, flavours or chemical additives; and animals providing organic meat are raised in accordance with humane standards. Meanwhile, COTA gives "local" a failing grade across the board. Unlike the term "organic" on food, which now, in many other parts of the world, is subject to government oversight, the term "local" can be slapped on any food with wild abandon. What does it mean? How local is local? Who checks up on whether a given item was really produced locally?

One potential comparative advantage that local has over organic is its generally smaller carbon footprint. Logically, if a product comes from the local vicinity, transporting it to your table should create fewer greenhouse-gas pollutants than the amount created when a similar item is brought in from abroad. This raises the question as to whether organic food is simply trading off one type of pollution for another: carcinogens for global-warming gases. Andre Leu, President of the International Federation of Organic Agriculture Movements (IFOAM), has thought a lot about this issue, not only because this Bonn, Germany–based global umbrella body for the organic sector represents 870 organizations in 120 countries, but also because he's an organic tropical fruit farmer from a remote area of Queensland, Australia. Shipping organic produce to buyers halfway around the world is what he does for a living. I met Leu at Natural Products Expo East in Baltimore. We sat for awhile chatting at the IFOAM booth on the convention floor, our conversation punctuated by a regular stream of convention goers approaching the booth to talk to Leu and peruse IFOAM's extensive stack of literature.

"When it comes to carbon dioxide, there are really two issues," Leu began. "In terms of the amount put into the atmosphere during the life cycle of a crop, the transport from the farm to the market is very minimal compared to other inputs. For instance, there's some wonderful work done in New Zealand: It's actually better for the environment for England to air-freight all its organic products from New Zealand every day as opposed to the way they're producing it conventionally now, particularly in winter when you have hot houses and heating." Leu's argument is that when you look at things like the amount of carbon dioxide and other greenhouse-gas pollutants like nitrous oxide that are given off by chemical fertilizers, along with the amount of carbon dioxide used to actually produce the pesticides, organic has a much lower carbon footprint than conventionally produced "local" food.

The other significant carbon advantage of organic, according to Leu, is the amount of carbon sequestered by organic agriculture in the soil. "The word 'organic' actually comes from the fact that we recycle organic matter in the soil through composting and the like," Leu said. "As a primary practice, our focus has been building up soil organic matter—in other words, organic soil carbon. And now we have very good data showing that organic practices not only emit less carbon in the growing than conventional, but also, when we factor in the amount of carbon dioxide we can strip out of the atmosphere and sequester in the soil when we build up the organic soil matter, we are actually not just greenhouse neutral. We mitigate more than we put out." Leu became more and more animated as he described new data indicating that a more widespread adoption of organic agriculture could have a huge impact on the reduction of greenhouse-gas pollution. Leu claims that with organic, we could mitigate up to 20 percent of all current greenhouse-gas output globally. Call this atmospheric detox.

I can think of no one who shows more passion for food issues than Sarah Elton. Sarah is a friend, a prominent journalist and author of the bestselling *Locavore: From Farmers' Fields to Rooftop Gardens—How Canadians Are Changing the Way We Eat.* She eschews packaged food, preferring instead to make all her family's meals from scratch, and she worries aloud that she wasn't able to can as many vegetables this past autumn as she usually does. I figured if anybody was prepared to eloquently and aggressively defend the moral superiority of local over organic it would be her. Turns out I was completely wrong.

For starters, in terms of her personal buying habits for her husband and two young daughters, she regularly shops at The Big Carrot and is "one hundred percent behind organic. We definitely need to be pushing in that direction," she told me emphatically. Her arguments are more aligned with Andre Leu's than I expected them to be. I told her about the COTA postcard and its criticisms of local food's lack of rigour or definition, and she agreed with it totally. "I was just chopping my pear this morning and remembering the sign about low-spray pesticides at the Brickworks [a farmer's market in downtown Toronto] where I bought it. What the heck does 'low spray' mean? Who decides what's low and what's high? What happens if you have pests one year? Are you going to tell me that you've gone high spray? I don't think so. So, I agree, local is not the be-all and end-all. It's a piece of the sustainable food system."

Elton defines her number one concern as "sustainable food," by which she means food that's good for people and good for the planet. She sees both organic and local as part of that equation. "There's no question that organic is better for the environment because organic agriculture nurtures the soil. If you're worried about carbon emissions, it sequesters more carbon than conventional agriculture. It doesn't rely on toxic chemicals because it sees agriculture as ecology, rather than a war against the earth. On the other hand, local food economies are so good on so many levels for rural and urban communities. So local plus organic equals sustainability: socially, culturally and environmentally."

Elton dismisses one of the more common criticisms of organic—namely, that it requires more land and results in lower yields and therefore cannot feed the world. She points to a recent study in the prestigious journal *Nature*, calculating that on average, organic yields are only 25 percent below those of conventional agriculture. "For me, this clearly shows that organic yields can compete. Think of all the money we've invested in conventional agriculture and the size of this industry. Meanwhile, organics is only 25 percent behind? Think of that. At the moment, organic yields are largely the product of grassroots people doing their own thing on their own farm. It's an amazing accomplishment. Scaling up is totally possible."[28]

Even though she figures her own family eats 80 to 90 percent organic, Elton isn't too fussed if, for the time being, local is stealing some of organic's buzz. "I see local as a first step in the learning process and the first step towards building a more sustainable food system. Hopefully, by realizing we want our local farmers, soon we'll be able to talk about why organic local is better than just local. We're losing our farmers so fast, so I'll take any farmer at any cost now, so down the road they might be able to be better."

In a nutshell, local food is organic food's gateway drug. It gets people hooked on thinking about where their food comes from—and then there's no going back.

Meanwhile, in California . . .

Perhaps the best indication that organics is coming into its own as a force in society is the coordinated and professional campaign mounted around Proposition 37 in California.

One of the myriad state ballot initiatives during the 2012 U.S. election, Prop 37 was designed to make the labelling of genetically engineered (GE) food products mandatory and disallow the use of the word "natural" on the labels of GE food. In order to get onto the November ballot, a massive grassroots organizational effort was undertaken by thousands of volunteers across the state. Over the course of 10 weeks, they gathered over a million signatures to force Prop 37 onto the ballot, nearly twice the number of signatures legally required.[29]

At the organic food convention in Baltimore, Prop 37 was the topic *du jour*. High-profile panel discussions were dedicated to it. "Yes to Prop 37" buttons were on display at every turn. People were giddy at the apparent lead enjoyed by Prop 37 advocates in the latest public opinion poll. And in typical fractious fashion, some organic advocates were busy publicly castigating those high-profile colleagues they deemed insufficiently supportive of Prop 37's passage.[30]

The communications efforts of the pro–Prop 37 "Right to Know" campaign were headed up by Stacy Malkan, the savvy co-founder of the Campaign for Safe Cosmetics and award-winning author of the book *Not Just a Pretty Face: The Ugly Side of the Beauty Industry*. Malkan told me that the Prop 37 campaign was important to her because, like her work on cleaning up the cosmetics industry, the Prop 37 initiative was largely a women's movement. It was about transparency and about taking back control over our lives from corporations that are wielding way too much power—in this case over the food we eat every day. Malkan called the campaign a watershed moment for organics: "an unprecedented show of muscle for the organics sector," with many organics companies "putting their money where their values are" and underwriting the Right to Know campaign's efforts. The public support for the campaign was also astonishing. According to Malkan, the campaign raised over US$2 million online from tens of thousands of smaller donors.

When I asked people in Baltimore why they felt so strongly about Prop 37, the answers I received almost always focused on the same issue: The promotion of genetically modified food is simply an excuse to spray more pesticides and keep the food supply under the control of corporate interests. Prop 37 is about democracy, they said. A recent study compared traditional versus genetically engineered (GE) crops[31] and found that farmers in the United States applied 318 million more pounds of pesticides over the last 13 years as a result of planting GE seeds (a.k.a. GMOs).[32] And increased pesticide spraying is not the only problem. As we've already described, Monsanto's new sweet corn seed has been genetically engineered to contain an insecticide in its own tissue.[33] In all respects GE food goes against everything organic food stands for. Malkan drew some parallels between food labelling and her efforts to ensure better labelling of cosmetics and safer cosmetics formulations. "What we have with GMOs is corporations modifying the genetic code of our most important food crops, with no proper health studies prior to these products being launched into the world. Consumers are being kept in the dark. The very least of what should be expected of these companies is that their GMOs be properly labelled so that consumers can decide for themselves what they do, and don't, put in their mouths." After years of dealing with the flawed science of major cosmetics companies, Malkan told me she was shocked at the extent to which corporations control the science surrounding GMOs. "Independent scientists are not able to validate the safety of GMOs because of the patents involved. The patent holders control the information: the same companies that told us DDT and Agent Orange were safe."

Though Prop 37 was ultimately rejected by California voters by a small margin (a few percentage points), Malkan is thrilled with the result. The opposition outspent her campaign five to one, carpet bombing California airwaves with their misleading advertising.[34] But the campaign's efforts brought the issue of GMOs to the forefront of American politics for the very first time and created a mobilized force on the ground that is ready for the next step.

In Malkan's mind, the Prop 37 campaign was just the beginning. In California consumers were clearly willing to fight for their right to an informed choice, and now state-level ballot initiatives are popping up in other places such as Washington state. Malkan points to the gay marriage issue, which lost repeatedly on ballot initiatives—and then one day it started winning. On election night in November 2012, gay marriage won on ballots in four states—in one fell swoop.

Nine Kids

In addition to Alex Lu's studies on pesticides and food, few other efforts have been made to evaluate whether organic food results in lower levels of toxins in your body.[35] So Bruce and I set out to see whether we could duplicate Lu's decade-old work. Because the task at hand for our big experiment was a pretty strange one—we needed a bunch of kids to agree to eat lots of vegetables over the course of 12 days during their summer break *and* they had to collect their pee each day. To find these kids, we did what any good 21st-century researcher does: We took to Facebook.

Within hours of posting a call for volunteers to all of my Facebook friends (I've got over 3,500 . . . ahem), we'd received interest from across the country to participate in our study. Our call outlined the following: The word on the street is that an organic diet might be better for the environment and our bodies, and eating organic could mean reduced exposure to the harmful chemicals that are in conventional food—something we wanted to show with their help.

We stuck to working with families who lived in the Greater Toronto Area. After a few days of interviews, we found nine really great kids from five families who were excited to participate in our experiment. Our final participants were made up of five girls and four boys, their ages ranging from 3 to 12 years old. These families were all different in a lot of wonderful ways, but they had two important things in common: They didn't at present eat any organic food, *and* they were willing to spend 12 days of their summer holidays helping out with our experiment.

Here's what the 12 days looked like for these nine kids:

Phase 1: Days 1–3:
8 a.m. collection of urine sample each day, conventional diet

Phase 2: Days 4–8:
8 a.m. collection of urine sample each day, organic diet only

Phase 3: Days 9–12:
8 a.m. collection of urine sample each day, conventional diet

We were very lucky to have that wonderful grocery store, The Big Carrot, offer to donate all the organic food for the organic portion of the experiment. And I'm not just talking apples and oranges—I'm talking organic apple sauce, organic zucchini bread and all that falls in between: olive oil, flour, canned tomatoes and, unanimously appreciated by our participant's mums and dads, organic ice cream for those hot summer days. So the night before the organic phase of their experiment, whether by car, streetcar or bike, each of these five families made the trek to The Carrot on Danforth Avenue to pick up their supplies for the next five days.

By now I'll bet a lot of you are wondering what happened with all that pee. The families kept each day's samples in their freezers until the end of the experiment, at which point our intrepid research assistant, Rachel, went around and collected the samples from each kid—all 108 jars of them. The urine samples were then shipped off to a lab in California for testing. And then we waited for the results. And waited. And waited.

A lot of anxiety built up over those months as we anticipated the correspondence from the lab. In this anxious state, receiving the results table with over 1,800 pieces of data on urinary pesticide levels was daunting, to say the least. It was quite a dramatic hour as we oh-so-carefully entered the lab results into a spreadsheet for our analysis, eager to see what they would demonstrate.

Our objective was to compare the organophosphate (OP) pesticides metabolite levels in the urine of the children. OP pesticides were chosen because of their widespread use, their reported presence as residues on foods frequently consumed by children and their acute toxicity.[36] Check out Figure 5 for a graph of the results (and the endnotes for an explanation of our calculations).[37]

Figure 5. Average of dimethyl dialkylphosphate (DAP) levels for nine children over the three phases of the experiment (units are ng/L)

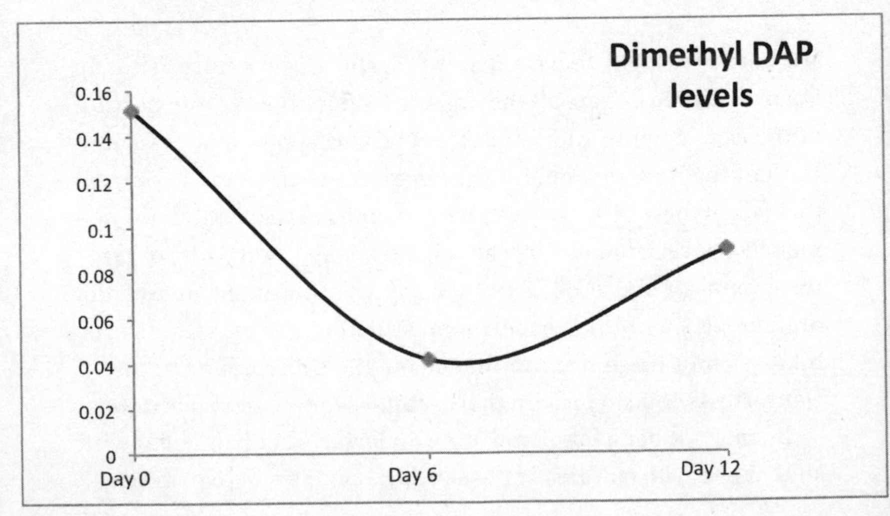

The graph, you'll see, shows average concentrations, for all the children together, of dimethyl dialkylphosphate (DAP) metabolites (which are common to OP pesticides) over each of the three phases of the experiment (conventional, organic, conventional). Notice the significant drop in dimethyl DAP levels during Phase 2 (the organic phase) and the increase again once the kids started eating conventional food in Phase 3. In scientific parlance, the organic and conventional results were "significantly" different. What does this mean to everyone in general? *Eating organic really can lower your pesticide levels!*

When we touched base with The Big Carrot to thank them and let them know about the exciting results, they were thrilled. Maureen Kirkpatrick, The Big Carrot's Standards Coordinator, told me she was delighted but not surprised. She noted that "these findings support what we at The Big Carrot know intuitively: healthy soil = healthy food = healthy bodies."

Another comment stuck with me when we did our follow-up interviews. One mother told me that she was initially interested in the study because her father, following some health issues, had become an "organic/vegan freak" but that she didn't really believe all the hype about organics and being chemical free. As a single parent, she was also wary of organics' cost. "Now," she said, half-laughing, half-sighing, "I've got some real numbers proving my dad is right." She said she didn't know what's more annoying: admitting that her father is right or that these chemicals are in her daughter at all.

I think we can all guess what she really believes.

The Doctor Is In

Pop Quiz: What do the following have in common?

- Canada makes seatbelts mandatory for all vehicle occupants.
- An increasing number of Canadian jurisdictions require helmets for child cyclists.

- Ontario becomes one of the first jurisdictions in the world to announce the closing of its aged coal-fired electricity-generating stations due to public health concerns related to smog.
- Over three-quarters of Canadians are protected from toxic lawn pesticides due to new municipal bylaws and provincial laws.
- Laws restricting smoking and cigarette packaging just keep getting tighter and tighter.

Answer? All of these policy advances were vigorously supported by the well-connected and well-resourced organizations representing Canadian doctors. Coincidence? I don't think so.

I first met Dr. Ted Boadway when he was Director of Health Policy at the Ontario Medical Association (OMA), a position he held for over 20 years. Throughout much of this time, he was also a family physician practising in Thornhill, Ontario. Boadway and I worked together on Canada's federal Chemicals Management Plan (CMP), the important policy framework that led to Canada's groundbreaking ban on BPA in baby bottles, and Bruce worked closely with Boadway on Ontario's coal-plant phase-out. Boadway ably served as Chair of the CMP's scientific advisory panel, and he told me he has always been interested in public policy issues, especially after a formative incident early in his career. In the mid-1970s the OMA produced a draft bill for seatbelt legislation, which, after a time, was actually passed into law. Boadway remembered wanting to get personally involved in the issue because he was doing some ICU work at the time in the small hospital where he was working, and the car-crash injuries he was seeing were just terrible. Boadway recalled a local Member of Provincial Parliament being invited to an OMA branch society meeting. "At one point in the meeting, he said to us, 'What would be the single thing the government could do to have an immediate impact on the health of people and maybe save the healthcare system some money?' And I gotta tell you the entire physicians' group said almost in unison, without any preplanning, '*Seatbelts*.' Seatbelt legislation on every lip. He was shocked, and he went back to his party and became a seatbelt campaigner." It was at that moment that Boadway realized the power that physicians have, if they choose to use it, to shape opinion. The public's opinion. And the politicans' too.

Why do physicians have this influence? According to Boadway it's simple. "What people are concerned about, and interested in, is their personal health. They're pretty confident that their physicians have their personal health at heart. So when physicians weigh in on matters, people realize it's their own health that's being talked about, and they start paying closer attention." Under Boadway's leadership, the OMA was careful in picking its battles. The association tried to choose areas where, first of all, there was good evidence of harm, secondly, where it was seeing damage, and thirdly, where people would feel the association was speaking in the best interests of their health. Abiding by these simple rules, time after time, the OMA won its arguments. Doctors got involved and won the argument on seatbelts. They were decisive in the battle for stricter drunk-driving laws. They won the argument on second-hand smoke. And against huge odds, they won the argument on the need to close down coal-fired electricity plants—which at one time were at the heart of Ontario's industrial economy—to reduce smog and protect public health.

A similar moment is approaching in relation to toxic chemicals. Already we see the American Medical Association (AMA) and Canadian Medical Association (CMA) expressing their concerns regarding chemicals such as triclosan.[38] The Ontario College of Family Physicians (OCFP) and the Canadian Cancer Society have become very active in seeking better controls on pesticides.[39] After interviewing many medical experts, the President's Cancer Panel concluded that the "true burden of environmentally induced cancers has been grossly underestimated." The panel advised President Obama to use the power of his office to remove the carcinogens and other toxins from our food, water, and air that needlessly increase healthcare costs, cripple our nation's productivity, and devastate American lives."[40] And for the first time in history, the American Academy of Pediatrics has ventured an opinion on organic food. Though not exactly a slam dunk for the organics industry, the statement does validate the worth of organic food when it comes to lower levels of pesticide residues and decreased risk of exposure to antibiotic-resistant bacteria.[41]

It's hard to believe now, but in 1976 Ted Boadway was yelled at by irate patients when he made the waiting room in his doctor's office a no-smoking area. He's not surprised that physicians' organizations are increasingly wading into the toxic chemical debate, given what he calls a veritable avalanche of scientific evidence being released every year. "Its time has come," he told me with confidence.

Boadway should know. When the doctors get involved, watch out.

THREE: STRAIGHT FLUSH

~ Bruce detoxes ~

Kick back in the back, get the phantom to drop
Bass blarin' outta my system, that's how I detox[1]

— AMY WINEHOUSE, "REHAB"

Walk Detox

IS DETOX THE "MASS DELUSION" that at least one doctor has described it to be?[2] Surely, something can be done to reduce our toxic burden and the various nasty diseases that may ensue from inadvertent exposure to, or ingestion of, the poisons around us. Take, for example, the chemicals that so vividly increased in our bodies when Rick and I experimented on ourselves in *Slow Death*.

As we have said earlier, an overarching objective of this book is to answer questions we received from readers, and in this respect I have given considerable thought to the many times we were asked: "Did you use any special detox treatments to lower your chemical levels *after* the *Slow Death* experiments?" The answer is no, so here is our opportunity to dig into that important question.

At a very basic level it is frustrating for so many of us to realize that despite all of our best efforts to avoid toxins in our food, our homes and our personal-care products, we are still exposed to hundreds, if not thousands, of potentially harmful chemicals on a daily basis. The answer isn't to cut ourselves off from the outside world as the occasional hardliner accuses us of advocating—

usually someone saying something like "What am I supposed to do? Live in a plastic bubble?"[3]

Step one is understanding detoxification. And to do that, in this chapter, we'll unpack and assess the multibillion-dollar detox industry[4] through the experiences of individuals and health experts who have devoted years of their lives to the complex, often expensive and sometimes painful journey of personal detoxification. Who better to learn from than these detox pioneers?

Detox Man

I first met Peter Sullivan following a San Francisco speaking event in Haight-Ashbury, a hippie hangout during the heyday of that phenomenon. Rick and I were on our book tour, and we had an amazing conversation with Peter over dinner following our talk. We were both amazed by his matter-of-fact approach to his detox obsession. (I use the word "obsession" with no ill intent but to describe accurately Peter's almost singular focus in life on trying to understand his body and the ways that toxic chemicals affect him and his family.)

For all of his life Peter suffered from various food allergies, as do so many of us these days. I happen to be allergic to pitted fresh fruit (peaches, cherries, etc.) and virtually all forms of pollen, mould and animals with fur. The allergy epidemic in industrialized countries is as mystifying as it is disturbing (for more on this topic, check out Chapter 5, where Rick discusses the link between allergies and phthalates). How many times has someone remarked, "In my day there weren't kids with lethal peanut allergies in every classroom. What's going on?" Exactly. What *is* going on? Peter wanted to tell his personal story about detoxing his life and the lives of his family members—including his devotion to self-experimenting with virtually every detox technique that exists.

In late spring of 2012 I returned to California and travelled to Silicon Valley, where I sat down with Peter to hear his tale of self-detoxification firsthand. I thought Rick and I were good at self-experimentation, but compared to Peter we are mere amateurs.

There was no possible way that I could test the hundreds of detox strategies that Peter has used over the past decade. As you will see, Peter is able to devote the necessary time and financial resources to this issue in a way that nobody else in the world has likely done. So though his work is well beyond the means of many people, there are lessons to be learned from a guy who has literally "done it all." Visiting with him not only helped me understand the stunning breadth and diversity of current detox techniques on the market; it also helped me identify what are probably the most effective techniques to use if—like Rick and me—you have more modest means.

Peter is an amazing individual with boundless enthusiasm, a boyish appearance and an ability to talk at great length (and I mean *great* length) about dozens of complex, interconnected ideas. He has a graduate degree in computer science from Stanford and designed and ran the web interface for Netscape. That may sound quaint now, but at the time it was the second-largest Internet portal in the world. Peter is also a genuine and easygoing guy who doesn't set himself apart from the rest of us. But the difference between him and the general populace who are concerned about the effect of synthetic chemicals on their health is this: He's spent the better part of 10 years of his life seeking answers to the basic question "What is going on?"

In addition to having allergies, in the late 1990s Peter began to face acute health issues. One night in 1998 he passed out and had difficulty waking up. Later, he suspected that episode had been caused by something to do with food, but he knew it was not food poisoning. He sensed he'd had some form of bodily imbalance. The fainting episode was the trigger for Peter's research into what was going on with his body and his life. He told me that he was irritable, stressed out, unhappy and feeling generally unwell. He was under tremendous stress at work and feeling burned out. Strange and unpleasant things were happening to his family too: One of his young sons was diagnosed with "sensory integration" and was experiencing emotional breakdowns, night tremors and autistic behaviours—including head banging. Peter's other son, at five years of age, was expelled from kindergarten for his antisocial behaviour. It was all too much for their father, and he figured there was more to what was going on than the everyday stresses of life, work and family.

Peter turned his sharp, analytical mind to the problems he and his family were facing and started to connect a few dots. Food was an obvious place to start. Were he and his family experiencing undiagnosed food allergies? Intolerances to lactose or gluten perhaps? Mercury also came to mind. Although at the time Peter was largely unfamiliar with the health problems associated with high levels of mercury, it didn't take too much digging around for him to discover that stress and irritability are among the many symptoms. If mercury was indeed part of the problem, what was the source of the mercury?

Fish and dental amalgams (mercury fillings) are the primary sources of mercury in most people. Peter ate a lot of fish and had many amalgam fillings. What's more, his dentist had informed him that he was grinding his teeth when he was sleeping—as many people do. He found the idea of grinding the mercury in his teeth and ingesting that highly poisonous neurotoxin (a toxin that affects the brain) to be literally sickening. And what about his kids? They didn't have mercury fillings that he knew of, but they ate fish and likely received vaccines with mercury in them (mercury is used as a preservative in vaccines). Peter began asking health professionals whether he needed to be concerned about mercury and its possible relation to his health problems and those of his kids—especially the autistic behaviours. One doctor had no advice to give, another indicated that mercury was not used in dental fillings and another said that ideas about links between mercury and the problems Peter's family was experiencing were "fringe." All these were typical of the responses traditional doctors would give—especially a dozen years ago.

But Peter did not give up. As a search engine designer, he did the obvious thing and went hunting for a new doctor online. He found a local doctor who specialized in "integrated medicine." One of the first things his new doctor did was to correct the misinformation regarding mercury fillings: Peter's mouth was *filled* with them. When he was tested for mercury, the results were shocking. His mercury level was 23 micrograms per millilitre—nearly three times higher than what is considered to be safe. He decided to have his sons and his wife tested too, and it turned out that they all had elevated mercury levels. This was the beginning of Peter's first dedicated foray into personal body detox.

He followed what is now a familiar procedure for heavy metal detox. Step one: he had his mercury fillings removed. This procedure reduces ongoing mercury exposure but also results in a mercury level "pulse," or rapid release into the body, so it needs to be undertaken with care. Step two: he embarked on chelation (pronounced "key-layshun") therapy. Put in simple terms chelation is a medical treatment that involves introducing specialized solutions into a patient, either orally or intravenously, which bind to heavy metals and cause them to be eliminated from the body via urine.[5] Peter was advised correctly that intravenous chelation therapy is the most effective procedure known to reduce heavy metal levels—and particularly mercury. I tried this treatment as well (and have described that process in detail later in this chapter).

Peter underwent two and a half years of intravenous chelation treatments with DMPS (2,3-dimercapto-1-propane-sulphonic acid), and he also tried chelation using DMSA (DL-2,3-dimercapto-succinic acid, the oral solution). At the end of the treatment period Peter felt better—his mind was clearer and he could think faster, although he still had problems sleeping.

We talked about the dozens of cleanses and diets that Peter tried as part of his detox quest. Rick made sure that I asked Peter about colonic irrigation, for instance. Peter said he'd tried it three or four times—the last time in Hawaii, where it made him feel quite good. Some of us go to Hawaii for the scuba diving and surfing; Peter goes for the colonic water sports. He cautioned me, however, that colonic irrigation can destroy the flora in the body. I managed to avoid the topic of fecal transplants, a bizarre but apparently effective means of replacing body flora, which is rather self-evident, given its name. Others go much further in their critique of the resurgence of colonic irrigation, stating that it is not only quackery but is associated with potentially serious health risks.[6]

Electromagnetic Personalities

After our lengthy conversation earlier in the day, Peter and I met for dinner in Palo Alto at a fine restaurant that serves local organic food and has an on-site drinking-water filtration system. We each savoured a glass of the water as though it were a fine California Zinfandel. Peter continued his tale from earlier that day, and we jumped through several topics, including ion balance, heavy metal toxicity, the effects of electromagnetic fields (EMFs) and his experience with chelation. He described his journey of moving through different levels of detox in the same way that someone might describe advancing through the different levels of karate. You could say that Peter has a "black belt" in detox.

One of the things that Peter is particularly preoccupied with is EMF radiation (not to be confused with the obscure British band EMF (Epsom Mad Funkers)). We know that many forms of radiation are important and necessary for life on earth: light and heat, for example. However, ultraviolet radiation from the sun, X-rays and naturally occurring radioactive elements, such as uranium, radium and radon, are able to cause harm to varying degrees. Electrical currents create electromagnetic fields, so any electrical device in our homes can emit EMF radiation, as do the increasing number of mobile phones, wireless modems, cellphone towers and, of course, microwave ovens. Remember the first microwave oven called the Amana Radar Range? A popular prize on 1970s game shows and, as with all microwave ovens, based on military RADAR technology. Electromagnetic radiation is invisible, and most people experience no symptoms during or after being exposed to them. But others find that certain EMFs are almost paralyzing. The ambiguity surrounding the effects of this type of radiation is only intensified by the fact that it seems as many studies have been carried out concerning the inconclusiveness of EMF research as on EMFs themselves. Going back to one of the basic tenets of risk—that is, where "risk of harm" equals "hazard" times "exposure"—the possibility of harm appears to be increasing because of the nature of the electronic devices we are using. But more importantly, exposure to EMFs is increasing exponentially due to the number of electronic devices that abound. At the end of 2011 there were 6 billion mobile phone subscriptions worldwide.[7] Six billion! With all of these mobile phones emitting direct, close-range EMF doses, risk by its very definition is increasing. If you are an avid mobile phone user you may have experienced the warming sensation on the side of your head after a lengthy phone call—this is caused by microwave radiation and can actually cause localised heating-up of your brain. In 2011, the International Agency for Research on Cancer classified mobile phones as possibly causing cancer in humans despite the fact that numerous studies on large numbers of mobile phone users have not found increases in cancer rates. Research and regulatory bodies are taking a precautionary approach in limiting the radiation levels of mobile phones and recommending that phone users, especially kids under 12, use hands-free features or text messages to keep the phones—and EMFs—away from their heads and brains.

My teenaged daughters, like tens of millions of others their age, appear to be fused to their cellphones and computers. Frankly, this situation horrifies me. Many healthcare professionals are worried about this combination of the massive global prevalence of new electronic and wireless devices and science that is unclear concerning the effects of EMFs. Some doctors stress that given the increasing evidence linking EMF exposure to poor health, it may be prudent to exercise caution in using the increasing amounts of wireless technology now available to us.[8]

After our tasty organic dinner, it was time to check out Peter's EMF-proof Faraday Cage, named after Michael Faraday, the 19th-century English scientist who discovered how electromagnetic fields work. I could see that getting to Peter's house was going to be half the fun: parked outside the restaurant was his white Tesla electric sports car. With some effort I squeezed my six-foot-two frame into the tiny passenger compartment, and off we sped out of Palo Alto into the misty hills of Los Altos. It was just past dusk as we zoomed around corners and over hills in the near-silent vehicle. The winding roads, lush tree canopy and high-pitched whine of the engine convinced me that at any moment we would veer off into a leafy driveway, through a retractable construction barrier and straight into the Bat Cave of my childhood TV memories. Not exactly, and yet not too far off, for we were, after all, en route to see the Faraday Cage, something I imagined The Joker could have designed to imprison Robin as bait for the caped crusader.

We arrived at Peter's house, the last on a dead-end road, halfway up a mountainside. I was given the quick tour. The home was a large, modern, three-storey structure, with fabulous views down the valley. Much of the structure was unused or under construction because of the extraordinary efforts Peter has undertaken to replace wiring to eliminate "dirty electricity," clean up mould, add high-end air filtration systems and remove toxic materials so that he can radically reduce exposure to synthetic chemicals and EMFs. ("Dirty electricity" is unwanted electromagnetic energy that can be produced by dimmer switches, compact fluorescent light bulbs or poor wiring. It is measured in Graham-Stetzer (GS) units.)

The Faraday Cage was what I had come for and Peter took me to an upstairs bedroom to check it out. Turns out the Faraday Cage *was* the bedroom. I had somehow imagined that there would be steel mesh involved, and that it would be a solitary cage sitting in the middle of a dark room; perhaps suspended from the ceiling in true Batman style. Ideally in a dank, windowless basement. But no, it was a pretty normal-looking bedroom with a small mattress on the floor, a desk and a table.

Peter took me to his office—a room down the hall—where a bizarre measuring instrument was sitting on his desk. It looked as if it had sprung out of Jules Verne's imagination or perhaps the *Wild Wild West* TV series. It resembled some sort of Victorian weather instrument made of glass and brass, attached to a metal box and featuring various switches and readout screens. A regular blip kept appearing on the screen, and I asked Peter what it was measuring. He explained that it was coming in from the rotating military radar perhaps five miles down the valley from where we stood. "And that is a bad thing?" I asked, knowing the answer. "Very bad," said Peter. Hence the Faraday Cage.

Having worked in software design and electronics all his life, Peter is now convinced that electrical devices are the cause of tremendous suffering for him, his family and millions of others. We walked into the Faraday bedroom, Peter carrying the EMF device, and sure enough, the radar signal ceased to blip. The "cage" was working.

It was nearly time to turn in and experience firsthand a night in the Faraday Cage. Before leaving, Peter wanted to do a quick check of the house to make sure there was no EMF contamination to disturb me. He grabbed the Graham-Stetzer STETZERiZER Microsurge meter and turned it on. This instrument measures "dirty electricity," and much to Peter's dismay, the readout showed a measurement of more than 100 Hz. Peter raced around the house, turning light switches on and off, trying to find the dirty electricity offender while I held the microsurge meter. "That's the one," I yelled out to Peter when I saw the readout suddenly plummet. The kitchen pot lights, very near the Faraday room, were the culprits. Peter was happy that we'd isolated the source but dismayed that the wiring in that location was still bad after all the time and money he'd spent trying to "clean up" his electricity.

Sleeping in Palo Alto was far more pleasant than being locked in a room in Toronto breathing stain-repellent perfluorinated compounds, as Rick and I did for *Slow Death*. I lay in bed in the Faraday Cage, handwriting my experience, free of the electromagnetic field my computer would have generated. The painted surfaces of the room concealed a coat of dark graphite paint, one of the main features of the Faraday Cage. Since it's a soft metal, the graphite provided a shield against the incoming radiation, as did the metallic film covering the windows. The entire room was also grounded. And as if that was not enough, at the foot of the mattress was a special sheet with grounding built into it and a ground wire going to one of the grounding points in the room. I started humming "Good Vibrations."

Peter loves to measure things; I do, too, but it's not often that I have so much information at my fingertips when I'm dozing off at night. With the various instruments at my disposal, I knew that I was 774 feet above sea level, the barometric pressure was 941 millibars, the temperature was 70.7 degrees Farenheit (21.8 degrees Celsius), the humidity was 55.3 percent, dirty electricity was 40 GS and EMF was almost nonexistent, at 5 Hz. Most of these measures seemed reasonable based on my observations, with the exception of the barometric pressure, which at 941 millibars was roughly equivalent to the pressure at the centre of a Category 4 hurricane. The light drizzle outside did not qualify. Oh, well. No need to worry, I was ready to sleep.

I wore a monitor on my wrist to record my sleep quality and download the data onto Peter's iPhone. That was the first thing Peter checked when he arrived in the morning. I had slept for 7 hours and 5 minutes. My sleep quality was 95 percent (that's technical jargon for "I slept like a baby"). Peter was impressed, but I told him anybody who knew me would not be. It's not uncommon for me to fall asleep on an airplane before everyone has boarded and to wake up when the wheels hit the ground at my destination. But according to the sleep "app," I'd had a good balance between light and deep sleep with almost equal amounts of each.

Peter's Purification Potions

Well rested and absent of electromagnetic contamination, I chatted with Peter some more, sitting on the sofa in his large, sparsely furnished living room, which looked out over the lush, fog-laden hills above Palo Alto. We delved into the various cleanses, supplements, waters and ionic cellular hydration drinks that he has tried out over the years. Peter fired off a zillion facts and figures, quoting dozens of books and health experts.

We shared a bottle (glass, of course) of water containing Zeta Aid crystals. "Helps the brain connect the dots," Peter said. He then handed me an antioxidant supplement to take with the water. I took one capsule of the Phyto5000 described as "a Powerful Anti Oxidant" with "an incredible 42,000 units of anti-oxidant power (or ORAC value) in 1 capsule." "It's to protect against free radical damage," Peter explained. Free radicals are those nasty molecules blamed for the majority of cellular damage that causes aging. Peter spoke of the many benefits of the supplements and cleanses we tested. Enhancing performance was a central idea, considering that heavy metals and other toxins may impair our ability to sleep and to think and function physically and that by ridding ourselves of these toxins, we can attain our peak performance. Energy, awareness and "the happiness feedback loop"[9] were among the factors that Peter connected to detoxification.

We discussed the "carrying capacity" of our bodies, a concept taken from ecology. Basically, our bodies can take only so much before something sends us across a threshold and our biological systems begin to collapse—the proverbial "straw that broke the camel's back." For some people it's the combined onslaught of pesticides, plastics, EMFs, heavy metals and air pollution that tosses them over the toxic cliff. And certain individuals do a "good" job of absorbing and retaining toxic chemicals, while others seem better equipped to eliminate and detoxify unwanted body contaminants naturally. It isn't clear why these variations occur other than knowing that individual body chemistries behave differently when interacting with chemicals.

I was curious to try firsthand some of the detox potions Peter told me about, especially after hearing how his health, and that of his sons, had improved so dramatically following their detoxification regime. Peter explained how his eldest son—the one expelled from kindergarten—was now excelling at school and no longer exhibiting antisocial behaviour. My plan was to spend the remainder of the day in Peter's house experimenting with various detox products. He rummaged through his kitchen cabinets, gathering dozens of boxes, packets and measurement tools, explaining how to use them. Peter wished me good luck and zoomed off in his Tesla, saying he'd be back at the end of the day. I used the handy litmus paper kit to test the pH of my saliva. It measured 6.3, or slightly acidic—I didn't know if that was a good thing or not, but I explain it below.

It was nearly noon and I was starting to feel a bit sluggish, plus I had an aching neck and head. Peter had said to me before he sped away that if I started to get a headache (I'd mentioned to him that I'm headache prone), I should take a swig of oxygenated water and "zap," he declared with a Zorro-like gesture, it would be gone. I was not really in the mood for more water, since I was heading to the washroom every 15 minutes. But I opened a bottle of the locally produced O2Cool and glugged a third of it down, then stopped and assessed the state of my head. Amazingly, it did feel better!

By early afternoon I had consumed nearly one hundred ounces of water (distilled or reverse osmosis, and some with Zeta Aid). This was the equivalent of 6 pints (a liquid measure I can most easily relate to, since my favourite barley beverage is served in pints). My energy level was dropping due to missing out on lunch, but I found an energy enhancer on the counter called Vison Present with "ionic cellular hydration," "nano-clustered." The instructions call for an eighth of a teaspoon of the fine white powder to be added to 8 ounces of water—just what I needed! The mere thought of it made me run to the bathroom one more time.

After a while I tested the pH of my saliva again and wonder of wonders, I had become a basic person. My pH reading had jumped five categories and now read 7.6. *What was that all about?* I asked myself. I did a little research and discovered that being basic (a pH above 7) is a good thing. In fact, a pH of 7.6 is considered to be in the optimal range (7.2 to 7.6) for good health. A cancer patient will sometimes have a pH of 3.5 or lower, and usually, the older you are, the lower your pH will be because of free radical damage associated with declining health. In fact, a pH below 6 is not ideal. Is it possible that my day-long consumption of Zeta Aid and other supplements took me into a pH zone of better health?

Peter picked me up at the end of the day, and after another delicious and much-needed dinner, delivered me to my hotel in Palo Alto. The next day I awoke in my no-doubt EMF-filled hotel room after another outstanding sleep. I wish I'd been wearing the sleep meter. I gazed into the lush, palm-tree-filled courtyard of my hotel, reflecting on my two days with Peter. He had used the image of a compass to demonstrate that in the end it was all about balance and that some of the detox products and routines seemed to work for him, notably chelation, hydration, avoiding EMFs and eating an organic diet, while others did not.

Let's be honest: Scientific proof that these remedies are making us healthier is decidedly elusive. We all crave definitive answers, but it's not always possible to apply general principles to each person's individual health. Health is, in many ways, an abstract idea, not really an entity in and of itself. It is defined by the absence of ill health. So it's best to pay attention to our own surroundings and lifestyle: what we do, what we eat and most important, what our bodies are telling us. There are some general factors, though, that apply to everyone. In essence, feeling good is about having energy, sleeping well and being in a positive frame of mind. Proper nutrition, hydration and exercise and adopting a non-toxic lifestyle are the backbone of good health.

Detox Docs: From "Quackery" to Convention

Armed with a reasonable understanding of detoxification, I felt it was now time to speak with some medical experts. Most of the research I came across was written by "health experts" but not by medical doctors. Nurses, homeopaths, naturopaths, health researchers or, in some cases, highly motivated patients, dominated the field. Much of their advice was no doubt valuable, but I admit I shared some of the biases that Peter Sullivan encountered when he was first consulting traditional doctors about his family's health. Shouldn't I be talking to "real" doctors? the ones who went to medical school?

I was also eager to speak with some MDs because the claims of success from detox treatments that I'd encountered so far were almost all anecdotal. There was no shortage of personal testimonials about the fantastic recoveries experienced by individuals, but there didn't seem to be many statistical studies evaluating detox treatments.

Ten years ago I doubt I would have found many detox experts who were physicians—doctors at that time were neither ideologically nor educationally inclined to accept that toxic chemicals at everyday exposures were a problem. This mindset is changing, however, and many doctors in North America now devote their medical practices to environmental medicine.

My first serious conversation about detox medicine was with Dr. Stephen Genuis, a well-known authority on the subject, based in Edmonton, Alberta. I wanted to learn more about his expertise in using different detox treatments. Detox medicine, Dr. Genuis said, is "in its infancy." He explained that medical testing is new and expensive and that it is difficult to measure chemicals because they move in and out of the body so rapidly. He also emphasized the fact that different people have different genetic makeups, including varying personal biochemistry. All of these factors make toxicological research and diagnostics challenging. He went on to explain that not only are people different, but the chemicals we are exposed to are different, and each one has its own nature.

How a particular chemical behaves depends on many variables, including the biochemistry of the individual. "So when I see a patient," Dr. Genuis said, "I'm trying to assess what toxins I think they have in them. Then I come up with a detoxification strategy or plan that is tailored to what I think they have." Dr. Genuis's approach addresses one of the main criticisms of detoxification: the fact that it is often not specific enough. Exactly which toxins are being targeted for reduction or elimination, and how are the before and after levels being measured?

Dr. Genuis noted that there are tens of thousands of chemicals in use, and the more toxic compounds that a person accumulates, the more those compounds will disrupt the biochemistry of their body. He first advises patients to find out where and how they are being exposed to toxins so they can adjust their habits, products or living situation. As he pointed out, "In order to get stuff out of people, you've got to stop putting it in." (This idea of shielding your body from further exposure is also the fundamental point of Rick's chapters and experiments in this book.) To identify the toxins entering his patients' bodies, Dr. Genuis conducts an inventory of potential exposure points with his patients. He goes through everything they are doing in their lives and then points out to them where exposure is occurring so they can make decisions about reducing or eliminating that exposure. This sounded a lot like the conclusion Rick and I came to in *Slow Death*: pay attention to everyday products and avoid the ones that are known to contain toxins.

According to Dr. Genuis, this step alone can be "life changing" for some of his patients. "When you remove what's going *into* the pail," he says, "then their body is able to devote its efforts to dealing with what's *left* in the pail." The idea that chemicals are not only harming our bodies—but are in fact preventing our bodies from detoxifying properly—is critical. If toxic chemicals are compromising our immune systems, our bodies are constantly fighting just to *stay* healthy and our major detox organs, such as our liver and kidneys, can't focus on their main job of removing toxins.

In order to understand individual biochemistry, Dr. Genuis typically runs a battery of tests on all of his patients. "I get the specific molecular makeup of their body so I see what they have and don't have [i.e., levels of essential vitamins, nutrients, heavy metals and toxins], so I can get their biochemistry to work properly." Toxic chemicals can disrupt our biochemistry and the more toxins a person bioaccumulates, the greater the chance that the toxins will disrupt our bodily functions.

Dr. Genuis is one of the few medical doctors who writes about clinical detox experiences in medical journals. He's also undertaken a detailed review of different detox therapies and has created a handy table that I have modified slightly for simplicity (see Table 6).

Table 6. Detox treatment evaluation

Type of Detox Therapy	Alleged Detox Mechanism	Detox Effectiveness
Fasting	Breakdown of body's fat cells and their stored chemicals, which are then released into circulation and available for excretion	*Limited* Some evidence showing caloric restriction can increase circulating concentration of some toxicants[27]
Chelation	Use of chelating agents (certain molecules and ions) known to chemically bind specific heavy metal toxins, which allows for excretion of toxic compounds	*Proven* Recognized treatment for some types of heavy metal poisoning[28]
Ionic foot baths	Feet placed in a bath, which sends a current into the body, generating positive ions that then attach to negatively charged toxins, which are excreted through foot pores	*None* No evidence in scientific literature. Clinical studies have confirmed no toxins are released.[29]

Type of Detox Therapy	Alleged Detox Mechanism	Detox Effectiveness
Colonic cleansing	Flushes encrusted material from the colon, which diminishes absorption of toxins and allows for improved excretion of body's waste	*None* Some research about benefits,[30] but empirical evidence lacking in scientific literature
Exercise	Breakdown of fat cells and their stored toxins, which are then excreted through lungs and perspiration and/or enhanced enzyme activity in detox pathways	*Limited* Some evidence confirming enhanced excretion of some toxic compounds through exercise.[31] (General health benefits of exercise make it important for overall health.)
Prebiotics and probiotics	Through dietary ingestion, restores damaged germ environment of intestines, which assists with gastrointestinal removal of certain toxic compounds	*Limited* Limited research associates prebiotics and probiotics with excretion of some toxic compounds[32]
Sauna therapy	Inducing perspiration results in toxins being excreted into sweat	*Proven* Some clinical studies have shown release of selected toxins[33]
Food and drink cleanses	Specific dietary interventions or restrictions will stimulate excretion of body's stored toxins	*None* No consistent confirmation found in scientific literature
Leeching	Leeches suck toxins from blood	*None* No confirmation in scientific literature

Type of Detox Therapy	Alleged Detox Mechanism	Detox Effectiveness
Herbal supplements	Certain supplements will facilitate excretion by enhancing body's natural detoxification processes	*Limited* Evidence suggests selected supplements may work,[34] but majority have not been confirmed in scientific literature
Hot springs	Immersing body in natural hot springs facilitates absorption of natural compounds that can enhance the excretion of toxins	*None* No confirmation in scientific literature of enhanced excretion through this method

Adapted from S. Genuis, "Elimination of Persistent Toxicants from the Human Body," *Human and Experimental Toxicology* 30 (2011): 3–18.[35]

Chemicals Get Stuck Inside Us

Not all chemicals are flushed out of our bodies quickly or fully. In fact, one of the properties that makes some of these chemicals particularly dangerous is their ability to attach themselves to various organs, fatty tissues and cells—or even to reside in our bones. This idea fascinates me, but it also scares me. Rick and I demonstrated in our last book that we could rapidly alter the levels of synthetic chemicals in our bodies, but I'm concerned that we may have been cavalier in our assumption that all the chemicals we absorbed or ingested left our bodies as quickly as they entered.

Heavy metals bind to proteins and tend to migrate into our major organs, such as our brain, liver, heart and kidneys. Among these heavy metals are mercury and lead, which are linked to heart and liver disease; cadmium, a potent neurotoxin; and lead, which builds up in our bones.

As if that weren't enough, Dr. Genuis explained that toxins can travel around inside us, moving from organ to organ thanks to our bodies' internal recycling mechanism, called enterohepatic circulation. Enterohepatic (intestine-liver) circulation essentially describes the flow of bile (and bile salts and acids) from our liver to our intestines. It is an efficient mechanism for reusing and recycling bile.

Our liver can remove toxic substances from our blood and discard them via bile into the intestines or the kidneys. These organs then eliminate the toxins from your body. But because bile recirculates through the body and because it is effective in carrying hormones and lipophilic chemicals (literally, "fat-loving" chemicals), it is one of the means by which harmful chemicals can recirculate through the blood and then the liver and be reabsorbed in fat or organs. Dr. Genuis maintains that this cycle can be damaging to our health because destructive chemicals don't remain stuck in one part of the body; they travel from organ to organ, finding a spot where they prefer to hang out. One of these chemicals might make its "home" in a fat cell in your brain, or it may become embedded in cells in your small intestine—and it's impossible to know what kind of damage it may cause once lodged in those areas.

The propensity of toxic chemicals to move around also explains why measuring levels in our bodies is not always straightforward. Doctors take samples from blood, urine, hair or feces and in some rare cases from sweat. Depending on the time of day, our drinking and eating habits, and other metabolic factors over which we have no control, a toxic substance that we are trying to measure may be more concentrated in the blood, liver, bile, urine, intestines or kidneys—or it may be stuck in a fat cell where we cannot measure it. This indicates that simple, "one-off" test results may not provide an accurate picture of toxicity, and it also explains why Dr. Genuis uses a variety of detox treatments with his patients in an effort to mobilize and remove toxins using all of our body's natural detox pathways.

Dr. Genuis has conducted what may be the first study of its kind that looks simultaneously at the levels of toxic chemicals in the blood, sweat and urine of the same individuals. His research demonstrates that we sweat out certain toxins, notably cadmium and lead, at levels many times greater than what is measured in our blood or urine. Individual body biochemistry, state of hydration and the location of stored toxins in our body may contribute to the differential excretion rates of toxins. Inducing sweating appears to be an effective means of eliminating certain toxic chemicals—particularly metals—that are present in our bodies. I'll put this idea to the test in the next chapter.

Dr. Genuis reminded me that the best approach to detox is to undergo a personal assessment that includes lifestyle, work history, diet, toxic chemical levels and an individual analysis of body chemistry. This assessment is particularly important for people suffering from one or more ailments, who are looking to detoxification as a prescribed medical therapy.

Synthetic chemicals are often lipophilic ("fat-loving," as just mentioned), and they include hundreds of cancer-causing agents such as some pesticides, polychlorinated biphenyls (PCBs) and compounds in household items like non-stick frying pans. They are notorious not just for being absorbed into fat cells, but also for building up in fatty tissue. That's why weight loss, as well as personal detoxification, is vital to the removal of toxic chemicals. As long as we're walking around carrying a chemical storehouse of blubber, we'll never adequately detox. The good news is that most detox treatments have the marvellous side effect of helping people lose weight: reducing fat limits the toxic storage capacity of our bodies.

Obesity is the number one killer in modern society. To quibble over whether the greatest cause of death is heart disease, stroke, diabetes or various fat-related cancers is to miss the elephant in the room, since they're all related to too much body fat. There's a fat epidemic among kids and adults, and it's increasing at an astonishing rate. Obesity does all sorts of nasty things to our arteries, heart, brain, blood, metabolism, mobility and more. And, of course, it provides the fat where toxic chemicals like to perch. "[These chemicals] . . . don't leave our bodies easily," Dr. Genuis said. "They deposit themselves in fat-loving organs, like the thyroid, the adrenal or the brain." Not only do different chemicals prefer different organs; Dr. Genuis said certain toxins may like to hang out in your flabby gut fat, while others prefer the fat in your butt.

Breasts, one of the fattiest parts of our bodies, are a primary repository for toxic chemicals—hence the association between synthetic chemicals and breast cancer. They "soak up pollution like a pair of soft sponges," says Florence Williams, author of *Breasts: A Natural and Unnatural History*.[10] Our bodies do their level best to isolate and eliminate unwanted toxins, and in this instance, the body releases the toxins in mother's milk. So through the most natural act of breast feeding, poisons are passed on to our newborn children, as Rick notes is also the case with phthalates.

Semper Fi: Always Faithful, a documentary film about male breast cancer, is an eye-opening exposé of the direct causal link between contaminated water, childhood exposure and breast cancer. For 30 years marines and their families living at Camp Lejeune, a U.S. Marine Corps base in North Carolina, were exposed to volatile organic compounds (VOCs) in their drinking water, which resulted in hugely elevated rates of male breast cancer, miscarriages among women who drank the water and many other diseases, including cancers.[11] Camp Lejeune is now infamous for having and allegedly hiding the worst case of drinking water contamination in American history.[12]

I had the opportunity to meet Mike Partain, one of the Camp Lejeune breast cancer survivors and a lead advocate for the affected families. Mike was not a typical candidate for breast cancer. He didn't drink or smoke, and there was no history of cancer in his family. In fact, he wasn't even a marine and had spent very little time at Camp Lejeune. But he was born on the base and lived the first few critical years of his life there as the son of a marine. The water that Mike drank as a child caused his breast cancer more than 30 years later. Were those chemicals stuck in Mike's body for all those years? Was there another trigger? perhaps hormonal changes resulting from his chemical exposure that caused his cancer to appear when he turned 39? Or were his cells damaged as a child, and it took three decades for the changes to manifest themselves in the form of breast cancer? These are important questions that need to be answered.

Mike has joined the outspoken advocates and survivors tired of being ignored by various authorities who have been pretending that chemical contamination is not a problem.[13] In August of 2012 they won a small victory when President Obama signed into law the Janey Ensminger Act, named after the daughter of a Camp Lejeune marine who died of cancer at age nine. The law provides medical coverage for the marines and their families who were poisoned at Camp Lejeune. But this was only a small remedy, and it came rather late.

Fear of cancer is one of the motivating factors behind the exponential growth in healthy living and detox programmes. We all know people who have been diagnosed with cancer. We know people who have survived and those who have succumbed. Over 40 percent of kids born in the United States today will be diagnosed with cancer at some point in their lifetime[14]—in short, nearly one out of every two children born today will be diagnosed with cancer. Stop and think about that. For every family of four, a family like mine, two out of four are statistically on track to develop cancer.

If that doesn't scare the crap out of us, I'm not sure what will. Perhaps the fact that over 550,000 Americans,[15] 75,000 Canadians[16] and 40,000 Australians[17] die of cancer every year. That's nearly 2,000 people dying every day in these three countries alone. Admittedly, cancer is still largely a disease afflicting older populations, but if half of us are going to develop cancer at some point, wouldn't it be worth it to figure out every possible way to prevent it?

We know that ingesting animal fats, eating a poor diet, smoking and not exercising enough may contribute to cancer, as well as to obesity, stroke and heart attack. And we know that many synthetic chemicals cause cancer. What you may not know is that for people who otherwise live identically, an exception in just one of these factors can cause cancer rates to vary significantly. That exception is diet. People who avoid all animal protein—that is, vegans—have significantly lower rates of cancer than omnivores or even vegetarians, who eat dairy.[18] Cancer isn't all about having an inherited disposition to the disease. By altering our diets, we can actually reduce our chances of developing cancer. And not just marginally: Vegans reduce their cancer risk by a third.

In *Food Rules*, Michael Pollan writes that nutrition science "is today approximately where surgery was in 1650."[19] He may be onto something, although I'd like to be a little more optimistic and suggest that nutrition science may be where surgery was in about 1850, not 1650. More books have likely been written on nutrition and detox in the past two years than all the books written on health in the years between 1650 and 1850. But that doesn't mean those books are necessarily filled with facts—as we will see.

The growing criticism of fad diets, and conflicting medical advice regarding diet and nutrition, risks undermining the entire world of nutrition and health. Are there common truths in the many detox books on the market?[20] Or are they really just long advertisements for diet products, as so often appears to be the case? One of the most frustrating aspects of the world of detox is that so many detox diets and cleanses don't even define or describe which toxins they are trying to eliminate. And most don't include methods of measuring the toxic chemicals entering and leaving our bodies.

How exactly do our bodies expel toxic chemicals? There are seven main pathways for detoxification: the lungs, liver, colon, kidneys, skin, blood and lymph system. In some cases these are simply the reverse routes of toxification. For example, we exhale toxic chemicals through our lungs in the same way that they can be inhaled. We ingest toxic chemicals in food and water and expel them as solid and liquid waste. And our largest organ, our skin, can play a helpful role in eliminating toxic waste, even as it absorbs the synthetic chemicals in lotions and creams.

The most important part of detoxification is to provide your body with the tools it needs to detoxify itself.

Livers, Kidneys and Lungs

Virtually all detox experts focus on our two main detox organs: the liver and the kidneys. Of those, the liver is our main detox workhorse, as well as being our largest internal organ. It's no coincidence that "live" and "liver" have the same etymological root: We cannot live more than a day without a functioning liver. I like to think of the word as a combination of "life" and "river," because in ecological terms, the liver is the headwaters of our body's detox system, flushing toxic chemicals through our various vascular and other tributaries, from which they're ultimately expelled.

The liver plays a dozen or more life-giving roles. It produces proteins and cholesterol, which help carry toxin-laden fat through the body. It manages our blood sugar by regulating the hormones insulin and glucagon, as well as performing important functions related to glycogen storage, regulating amino acids and proteins, producing immune factors and scrubbing toxic chemicals and drugs from our bloodstream. The liver is a temporary storage depot for blood, sugar, iron, vitamins and even the toxins that it is attempting to remove. In addition to filtering our blood, the liver manufactures bile, which helps carry away toxic waste. The liver and kidneys process and send toxins through the body's two primary waste pathways, leading to the ultimate elimination of waste products through the bladder and colon. Our liver is the organ that "decides" what ends up in those waste streams.

The first detox process that our body manages is the liver–kidney–bladder connection. Blood produced in the liver is filtered through the kidneys, where impurities and excess minerals are removed and carried away by urine, which is manufactured by the kidneys. Urine is stored in the bladder prior to being expelled. It is this highly efficient toxic waste elimination process that Rick and I rely on when we test the levels of toxic substances in urine.

The second elimination pathway is the liver–bile–gallbladder–intestine connection, which can be described as the highway of fat. If you've ever wondered where all that excess grease and fat from those yummy French fries ends up, here's the answer: it goes to the liver, which sends it to the bile, where it's broken down and then sent through the gallbladder to the intestines. It's digested further in the intestines before being expelled by the colon.

Suddenly, understanding this process starts to provide explanations for some of the many health problems plaguing the populations of industrial nations. We consume too much fat, making it impossible for our liver and gallbladder to keep up, and this results in fatty residues entering our intestines that have not been completely emulsified. (Emulsification is the process of combining oil-based and water-based liquids to form a consistent mixture, in the way egg yolk and oil combine to make mayonnaise.) Add to that the lack of fibre in most Western diets and we are further impairing the ability of our intestines to digest, process and eliminate fat and waste. We also know that so many of the toxic substances we need to avoid are fat soluble—and are therefore stored in the fat that our body can no longer process effectively. As if that wasn't enough, the triple threat of poor diet (excessive fat and lack of fibre), pervasive toxic chemicals and lack of exercise means that our livers have even *more* difficulty doing their primary job of ridding the body of toxins. The more toxic chemicals we ingest, the more we reduce our body's natural cleansing ability. Is it any wonder that we have a nightmarish epidemic of fatty livers, high cholesterol, diabetes, gallstones, liver cancer, bladder cancer, colon cancer and generally poor health? The importance of a well-functioning waste elimination system cannot be overstated. Colorectal cancer alone is the second leading cause of cancer death in North America.[21]

How is it that breathing air can result in toxic chemicals damaging our heart, our liver or a developing foetus? The primary function of our lungs is twofold, of course: taking oxygen from the air and transferring it to our blood and taking carbon dioxide from our blood and releasing it to the atmosphere (known, respectively, as inhaling and exhaling). When we inhale polluted air, however—for example, air contaminated by the off-gassing plastics in the interior of Rick's car or by a phthalate-filled home air freshener—our lungs transport not only oxygen to our blood, but also the toxic chemicals we breathe in. Anytime that we can detect a strong chemical smell—think turpentine or gasoline—we know we are inhaling the toxic chemicals in those products. Often, the stronger the smell, the more volatile the chemical and the more damaging the effect on the body.

Though it's generally well known that the lung filters out particles before they enter the bloodstream, its other detox functions are more obscure. However, there's one prominent volatile organic compound that is readily expelled through our lungs—alcohol. In fact, we can use the alcohol levels in our breath to estimate exactly how much is in our blood. It makes me wonder how we so readily accept the court-tested precision of a Breathalyzer in measuring blood-alcohol levels while industry and government agencies still question the validity of measuring toxic chemicals directly in our blood or urine. Rather than using nasty and invasive blood, sweat and urine sampling techniques in our self-experimentation, could Rick and I have just blown into some device to measure the levels of toxic chemicals in our bodies?

Using our breath to analyze our health is not too farfetched a notion, and it's far from new. Hippocrates, the Ancient Greek father of Western medicine, recognized that we could use breath to detect health problems,[22] and now breath tests are being used to detect cancer. Researchers have also found that chemicals absorbed by our skin can be found in our breath, and in some cases chemicals or medications used in the past can be detected in the breath many weeks later. This reinforces the understanding that exhaling is another important mechanism our bodies use to get rid of unwanted toxins.[23]

After some digging I discovered that doctors are already measuring toxins in our breath as part of their environmental testing—and that they can detect not only toxins, but also female birth control drugs in the breath of . . . men. A substance that is manufactured into a pill, ingested by a woman, eliminated through her urine, processed in a water treatment facility, dumped into a lake or ocean and then consumed in drinking water can show up in the breath of anyone who drinks that water.[24]

Breathing, sweating, peeing and pooing are the natural processes that flush toxins from our bodies. For the most part these mechanisms do a good job. So we'd be well advised to keep them healthy and functioning optimally, especially given the growing number of chemicals we take in every day.

Chelate This

It was time to test one of Dr. Genuis's "proven" detox therapies firsthand. I hopped a plane to Houston, Texas, to visit Dr. Peter Erickson, my first cousin and well-established environmental physician, and a specialist in chelation therapy. Dr. Erickson greeted me, along with his wife and their two Yorkshire terriers. We travelled the 90 minutes from the airport to their house in rural Texas—through Houston's drab, sprawling suburbs and on to the cotton fields, ranches and pecan plantations surrounding the bungalow on their small organic ranch.

I wasn't looking forward to having my arm poked with a needle and lying around for three hours with an intravenous solution of chemicals coursing through my veins. But that is what chelation therapy is all about, and it's the most specific and well-researched detox treatment available—and the one with the clearest demonstrated effectiveness, particularly for heavy metal detoxification.

Doctors use chelation to treat everything from arthritis to arteriosclerosis (hardening of the arteries) to autism associated with mercury to fibromyalgia (a condition characterized by chronic pain and fatigue, with unknown cause) and Parkinson's disease. The use of chelation as an alternative to heart bypass surgery is truly amazing.[25] But chelation is not without controversy. The National Institutes of Health (NIH), an agency of the U.S. Department of Health and Human Services, halted two major clinical trials to assess the benefits of chelation (one on heart disease and one on autism) after physicians raised fears that the procedure was not supported by enough scientific evidence and the study could put participants at risk. The opposing doctors indicated that, partly because there was no clinical evidence that chelation was beneficial, this first-ever study designed to *assess* clinical evidence should be stopped. The doctors supporting the research prevailed and the studies continued with the more than one thousand registered participants and concluded in 2012, but the outcome of the study continues to divide doctors. The formal results showed that chelation (as opposed to a placebo) offers modest benefits for recovering heart patients.[26]

To be clear, there is no disputing the fact that chelation is an effective treatment for heavy metal toxicity, including toxicity induced by lead and mercury. Doubts about the effectiveness of this therapy relate largely to the contentious connection between mercury and autism and therefore the use of chelation as a treatment for autism. There have been cases where chelating agents have been too aggressive in the removal of essential elements, especially calcium, resulting in a number of deaths due to heart failure in the United States. Calcium in blood plays a critical role in maintaining a heartbeat.

Given that chelation is an invasive medical treatment and not without risk of complications, only trained medical doctors should administer intravenous chelation. This is a requirement in some jurisdictions but not all. Chelation is essentially an application of chemistry, so I felt some additional relief that Dr. Erickson has a degree in chemical engineering on top of his medical degree.

Lazy Boys and Dogs

The chelation therapy I received can best be described as a "demonstration project." My tuna-eating experience in *Slow Death* showed quite vividly that the mercury from the tuna I ate was piling up in my blood at a rapid rate, nearly tripling in 48 hours. The goal of the chelation therapy was to do a "reverse tuna" experiment: I wanted to find out whether chelation could remove mercury from my body and whether I could measure the mercury actually leaving my body. I wasn't convinced it would be possible with only one session.

The experiment was divided into two parts. Day one was called pre-provocation, meaning that I would undergo a baseline assessment of how much mercury I peed out in a normal day. Day two was treatment day, when the chelation chemicals scavenged my body for heavy metals, literally grabbing onto the toxic molecules and removing them via my kidneys and urine. The term "chelate" comes from the Greek word for lobster claw (*chela*), referring to the pincer-like action of the chelation chemicals.

When the big day arrived, I was more than a little anxious. After I ate a bowl of organic fruit, Dr. Erickson and I headed into the small, dusty town of Boling, where his medical clinic is housed. The roadsides appeared to have remnants of freshly ploughed snow, but I knew that was impossible, given the temperature. Upon closer inspection I realized I was looking at lines of cotton left behind from the harvest that had just taken place. Boling is not much more than a very wide intersection with a few old buildings, a gas station, several large trucks and a couple of stray dogs, who met us at the clinic. Dr. Erickson pointed out that the slightly sagging clapboard clinic was itself one of the oldest buildings in Boling and had housed the town's first doctor's office and pharmacy.

I sat down in a large, reclining chair for what I considered to be the worst part of the experience, having the IV tube inserted into my arm. I explained my dislike of needles and, sure enough, poke—nothing. Poke again—nothing. Dr. Erickson looked at me, smiling, and said, "Bruce, you aren't going to believe this, but I can't seem to find your vein, and I'm not going to try a third time." At this point Suzy the nurse stepped in, mumbling something about doctors not knowing what they are doing, and after applying a hot pad to get my blood flowing, she found my vein easily and we were off to the races.

Chelation is the opposite of most medical experiences. It is a peaceful, relaxing, unobtrusive procedure (apart from the IV), where you basically sit back in a comfy chair, feet up, reading, listening to music and letting the little chelation lobsters run around your bloodstream catching their toxic metal prey. One of Dr. Erickson's Yorkshire terriers decided to keep me company and lay on my lap for most of the time that the needle was in my arm. I'm convinced it sensed the need to provide some reassurance. During both the pre-chelation (day one) and the chelation procedure (day two), I consumed large quantities of water, which allowed for the collection of nearly four litres of urine over each of the two six-hour urine collection periods.

After the three-hour chelation session was over, Dr. Erickson and I headed into the largest nearby town to pick up some organic veggies for our dinner. We stopped in at Starbucks for a tea (and for me to pee in my now rather heavy and obvious jug). This reminded Dr. Erickson of another detox treatment called a coffee enema. I made the mistake of asking what that was, and the answer was all too evident.

Back in Toronto several weeks later, a large envelope arrived by courier. I tore open the package from Doctors Data Inc. (a lab in St. Charles, Illinois), anxious to see my test results, and plain as day I could see that chelation had done an amazing job of removing toxic metals from my body. The graph showed a striking contrast between the amounts of aluminum, lead, mercury and tin that were removed from my body during chelation and the amounts that left during pre-provocation (my normal state) (see Figure 6). More than 4 times as much aluminum was pulled from my body during chelation, over 5 times as much lead and mercury, and 20 times the tin. The first three metals are serious poisons, and it's not safe to have *any* amount of lead or mercury in our bodies.

Figure 6. Bruce's urinary heavy metals: levels of removal before and during chelation (ng/g)

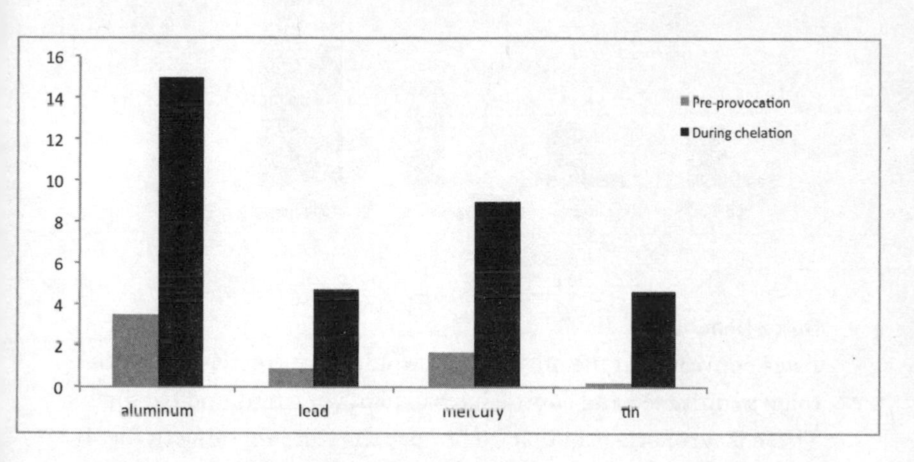

The chelation procedure had a similar effect on fundamental elements in my body, such as sodium, calcium and magnesium, with between 2 and 15 times the level of removal while I was undergoing the intravenous treatment (see Figure 7). This not only demonstrates the effectiveness of chelation in this case, but also points to one of the risks that must be managed. Unlike the poisonous metals that the chelation targets, sodium, calcium and magnesium are essential elements that our bodies require. Calcium, for example, is a metallic element that plays a critical role in the electrical impulse that keeps the heart beating. Flushing the mercury and lead from my body was a good thing, but chelation requires careful monitoring to ensure that critical elements are not depleted.

Figure 7. Bruce's urinary essential elements: levels of removal before and during chelation (ng/g)

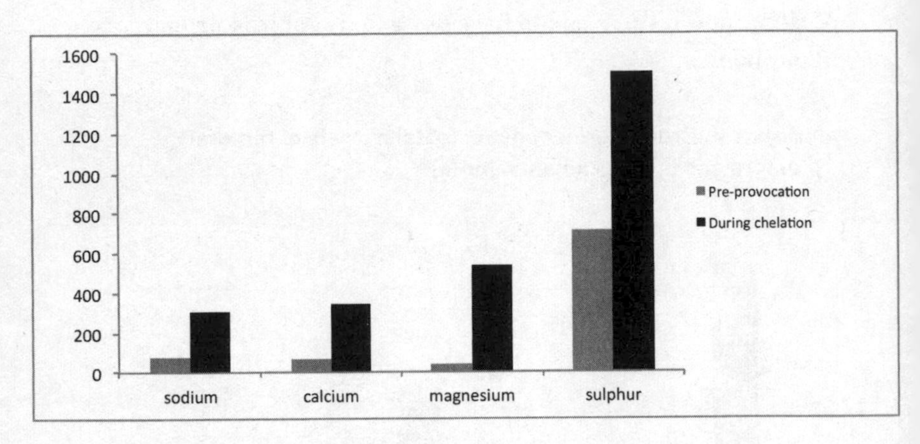

Detox Debunked

I was convinced at the outset that I would find the field of "detox" to be a conflicted and contrarian world of wing nuts and fad diets. There is no shortage of that, to be sure, but like anything in life, if you dig a little deeper, sometimes you find nuggets of truth.

Thousands of people are ready to take our money by selling us detox products, thus creating the global multibillion-dollar detox industry. Some remedies are helpful and some are hokum, and certain elements of the detox industry have figured out how to prey on people who are unwell and vulnerable. To defend yourself against the iffy or entirely useless treatments, take another look at Dr. Genuis's advice, summarized in Table 6. There is no point in paying for treatments that don't work, and specialized therapies, such as chelation, are meant for people with serious, medically diagnosed toxicity.

On the other hand, practitioners of conventional medicine are often too quick to dismiss the potential benefits of various approaches to detox, and the doctors who are most skeptical end up chasing their patients away and into the world of alternative medicine, where they find people who are willing to listen and offer solutions. We need to seek a healthy middle ground.

Of course, it's impossible to prevent disease entirely, so we need to maintain a significant capacity for treatment and cure. But we need to start bringing down the curves representing the incidence of toxin-induced disease, not just the graphs representing cure rates. This means shifting the emphasis so that health expenditures and cancer research dollars reflect the medical and ethical importance of preventing disease. Alternative medicine and personalized detox treatments can play a critical role in this shift.

We are all full of toxins. There's no question about that. The issue is how to figure out the most effective course of action for each individual, based on their specific health circumstances. Understanding how our bodies work so we can detox naturally is the first step in evaluating which detox treatments to use. We were not designed to process and eliminate modern synthetic chemicals, so it is understandable that our bodies may have trouble removing pesticides, volatile organic compounds and perfluorinated compounds, to name a few. First and foremost, we must avoid toxic chemicals in our food, homes, workplaces and lives in general. If no toxins, we don't need toxouts!

Most of us aren't going to spend many of our waking hours figuring out how to detox, nor are we going to build Faraday Cages or undergo chelation. So what about the easier and cheaper detox approaches? Herbal formulas, supplements and all of those off-the-shelf cleanse diets? That's the subject of the next chapter.

FOUR: SWEAT THE SMALL STUFF

~ Bruce perspires heavily ~

Don't sweat the petty things and don't pet the sweaty things.

—ANONYMOUS

Potions, Pills and Detox Diets

I AM GENERALLY VERY HEALTHY and don't get sick often. I exercise modestly (though I need to do more) and eat a healthy diet (though I need to eat less). So entering the detox world for me is very different than for many who are suffering from serious, sometimes debilitating, symptoms for which they are not finding remedies through conventional medicine.

Once you immerse yourself in the detox world, it's a hypochondriac's dream come true. I'd never thought about the pH of my urine or wondered whether my enterohepatic cycle was functioning properly. I've never been concerned about whether or not my blood cells were well hydrated. Or whether my fat cells were filled with organophosphates. And what about all that fat anyway? What do you mean I'm overweight? Surely I can just sweat out a few pounds and not do anything drastic.

I really wanted to dig into the debunking side of things, and I asked my friend Bryce Wylde, who is a homeopath and author, where he stood on the detox industry. "Craziness," he said. "You get people running out and taking things because the Kardashians are." In terms of the natural products on the market, he maintains that "50 percent are junk and 50 percent are hidden gems worthy of further research." Two of these gems are fatty acids and vitamin D—natural supplements with tremendous, well-researched, proven and medically accepted benefits. "There are lots more of those out there," he said. We just need to continue exploring and researching them.

My Google search of "detox diet" turned up over 5.8 million references, including everything from relatively reasonable approaches to nutrition and lifestyle to thousands of cleanses and diets of questionable repute. If you want to play a fun game, Google "detox diet" in quotations, then type in your favourite fruit or vegetable and see what happens. The lemon detox diet, the grapefruit detox diet, the avocado detox diet, the African mango detox diet, kiwi detox recipes, to name a few.

What seems to happen with diet fads is that somebody takes a basic idea—eat more fruit—and turns it into a gimmick, which they then profit from. So rather than focusing on fresh fruit as part of a healthy lifestyle, we end up with the "insert favourite fruit here" detox diet.

Clearly, one of the downsides of the Internet age is that anyone with a computer can create a diet, design a web page and make stuff up. Take one "lipid detoxification diet," for example: The recipes include eggnog, potatoes fried with bacon, steak ("eat as much of the fat as you can") and whipped cream. The recipe for a baked potato suggests that it be eaten with "at least half a stick of butter and a quarter cup of sour cream."[1] Wow, now that is a detox diet I could live with! Or maybe die by! And judging by the ever-increasing girth of North Americans, perhaps there are a few too many people who have discovered the lipid detoxification diet, whether they know it or not. My advice, if you happen to be one of them: It's time to switch!

As the number of detox products and treatments grows, so does the number of critics who emphasize the lack of scientific evidence that these methods actually work. The Australian consumer products watchdog group Choice reviewed seven detox kits available at pharmacies and health food stores. Their conclusion, based on their expert panel review, was that most of the detox programmes (typically 10- to 15-day special diets) were of little or no value—and one of the detox diets they tested was considered to be dangerous due to the low levels of protein and nutrients a dieter was allowed to eat during the course of the regime.[2] Other studies confirm the lack of evidence regarding the benefits of detox and cleansing diets.[3]

There are three main reasons why most detox kits are of little value:

1. They provide individuals with a false sense that consuming juice or various herbs over 10 days can somehow eliminate toxins that have accumulated over a much longer period of time. They therefore ignore one of the foundations of the detox process—namely, that detox is a continuous lifestyle shift, not a short-term diet.
2. They often recommend accompanying diets that restrict the intake of protein and other nutrients to such an extent that people unfamiliar with basic nutrition may suffer.
3. The most restrictive diets can lead to rapid and unsustainable weight loss, and much of that weight is regained after the cleanse. This can lead to unhealthy "yo-yo" dieting.

Detox kits do have some positive attributes, though. First, detox kits may be an "appetizer" for people who want to begin following a healthy lifestyle. We all need support mechanisms and tools to guide us. Using a detox kit may be the first step that someone takes on a longer pathway toward a sustainable detox lifestyle. Second, virtually all cleanses recommend avoiding junk food, caffeine, tobacco and alcohol—all of which do need to be ingested in moderation or not at all.

If a person typically drinks four cups of coffee a day, eats a doughnut in the morning, a burger, fries and a Coke for lunch, followed by a dinner of canned spaghetti—they will almost certainly feel better after two weeks of water, vegetables, legumes and a high-fibre diet. But for a diet like that, they won't need the detox kit.

I asked Wylde what the future holds for detox medicine, and again he gave an instant, specific answer: "Genomic diagnostics." Huh? "Genomics are used to evaluate personal impairment and our ability to detoxify naturally," he explained. Genomics is the field of measuring and evaluating the genome, and the genome is the complete DNA *sequence* that determines who we are. It is distinguished from genetic research, which focuses on individual genes, not the entire genome. Genomic diagnostic tools combine what researchers and medical practitioners know about genomic and clinical data to better understand the basis of disease to assist treatment and prevention.[4]

Genetic research into individual genes has helped identify those associated with diseases like breast and ovarian cancer. In the spring of 2013 Angelina Jolie made headlines, not as the lead in a summer blockbuster, but because of her decision to have an elective double mastectomy to reduce her chances of developing breast cancer. Jolie has a mutation in the Breast Cancer 1, or BRCA1, gene—the same mutation her mother and aunt had. Jolie's mother died of ovarian cancer at the age of 56. Her aunt succumbed to breast cancer just two weeks after Jolie's surgery. She was 61.

The BRCA1 gene plays a critical role in fighting cancer. If the gene is damaged, it fails to protect the person with that gene, leaving them with a 65 percent lifetime risk of developing breast cancer.[5] By choosing a proactive double mastectomy, Jolie has reduced her chances of developing breast cancer to under 5 percent.[6] Her decision was important, given her genetic predisposition to cancer.

However, 90 to 95 percent of breast cancer cases cannot be attributed to BRCA1 or other inherited gene mutations.[7]

There's a need for more research into environmental exposures that contribute to cancer, and the emerging field of epigenetics is doing just that. Epigenetics is the study of "gene expression"—how genes can be turned on or off by external, non-genetic influences—and it is at the heart of endocrine disruption research. Recent studies show, for example, the epigenetic effects of bisphenol A (BPA) on reproductive function. Male rats exposed to BPA during the first five days after their birth had impaired fertility when mature.[8] The research demonstrates how developmental exposure to endocrine disrupting chemicals can alter gene function much later in life.

The point is that different people react differently to environmental stress because our bodies all have unique features, and we cannot dismiss the extent or nature of these differences. Epigenetics therefore reveals that our genes themselves may not be the determinants of cancer or other health problems. It's how our genes express themselves and the extent to which synthetic toxins play a role in switching gene functions on or off that may trigger genetic disruption or disease, such as cancer.

And now, back to genomic diagnostics, the tool that Bryce Wylde described. "With genomics," he told me, "we can identify individuals who are more likely to have a negative reaction to toxins based on SNP factors." SNPs (pronounced "snips") are single nucleotide polymorphisms in our DNA that may, for example, predispose us to disease or influence our response to a drug. Bryce used the example of certain patients who may have serious negative reactions to chemotherapy because their bodies are not well adapted to detoxifying harsh, cancer-fighting chemicals. Others, however, may be so good at detoxifying that a standard chemotherapy dose may be ineffective in treating their cancer.

Genomics and SNPs may therefore be able to tell us which person has a predisposition to detoxifying and which one does not, so the proper treatment can be administered. We would also be able to identify people who have heightened chemical sensitivity. And this would be a great advance in the effectiveness of individualized, preventive medicine.

If genomics really delivers, we'll be able to discern, with laser accuracy, which people need to be examined and which need to be treated and precisely which therapies are best suited to an individual's genetic makeup. This approach might also explain why apparently healthy people exposed to minute quantities of certain toxins end up with cancer, while others who smoke two packs of cigarettes a day live to be 80 and do not develop the disease. We're not there yet, though, so for now we'll have to do our best by using effective methods of contemporary detox medicine.

Free Radicals, Antioxidant Rebels and Natural Herbals

Dr. Stephen Genuis, the detox medicine expert based in Edmonton, Alberta (whom I mentioned in Chapter 3), discussed one detox treatment that has some merit: supplements.

The world of vitamins and herbal supplements is huge, with global sales estimated at over US$68 billion. When I was young, Linus Pauling, the father of orthomolecular medicine (an alternative practice that focuses on achieving health through nutritional supplementation to compensate for inadequate dietary nutrition) and perhaps the best-known chemist of the 20th century, published a number of controversial books on the benefits of taking high doses of vitamin C. Pauling took 10,000 mg per day of vitamin C and up to 40,000 mg when he wasn't feeling well. The latter dose is four hundred times more than what is typically recommended as a daily dose.

Although Pauling was the only person who has ever won two unshared Nobel Prizes, he was discredited by the medical community for his claims that such high doses could help prevent colds and fight cancer.

I was a vitamin skeptic myself before I began this research, but I'm now convinced that a vitamin C supplement of at least 250 mg per day is quite likely beneficial for its antioxidant and generic detox properties. This amount is one-fortieth of what Linus Pauling took daily, but anything more than 250 mg is now thought to be of questionable value because vitamin C is water soluble and large doses are therefore easily removed in urine. Pauling may not have got the doses right, but a growing number of conventional medical doctors believe Pauling was onto something, and orthomolecular medicine is seeing a resurgence.[9]

Vitamins A, D, E and K are also important for detoxification, but they need to be taken with greater caution because they are fat soluble, and high doses may result in these vitamins lodging themselves in fatty tissues to dangerous levels, in the same way that toxic chemicals do. If there was one simple health supplement that virtually all people would benefit from, it is vitamin D. Throughout the globe, people are deficient in vitamin D—especially those who spend a lot of time indoors, or in northern regions, and therefore do not receive enough natural vitamin D from sunlight.

We should be able to meet our minimum daily requirements of most minerals and vitamins (other than vitamin D) by eating sufficient quantities of organic fruits, vegetables and whole grains. But it's a practical challenge to measure our intake and to know that we are meeting our daily requirements. That's why it's smart to take a multivitamin every day that delivers minimum requirements of most essential nutrients.

When I asked what he sees as the central health issue that we're facing today, without hesitation Bryce Wylde said, "Free radicals." Free radicals aren't Greenpeace volunteers; they're atoms or molecules that contain unpaired electrons. This explanation may bring back horrible memories of high school chemistry (or happy ones for nerds like me), but it's important to be aware of the role that free radicals play to better understand detoxification and good health. Free radicals are basically unstable, itinerant atoms that rove through our bodies, hooking up with unpaired molecules and causing indiscriminate damage to our cells. "They are without doubt the most gnarly and dangerous toxins inside us," Wylde stated, as though he held a personal grudge against them.

Free radicals are charged atoms that are highly reactive. They steal electrons from other molecules in our cells, a process called oxidation, which can trigger chain reactions resulting in rapid and extensive cellular damage—even to our DNA. Free radicals are blamed for chronic diseases that appear as a result of aging. Chief among these diseases are heart disease, Alzheimer's disease and cancer. A simple way to picture how free radicals cause damage to our cells is to imagine a rusted-out car. When iron oxidizes, it turns into rust, and rust is what ultimately destroys iron. This is why Dr. Oz refers to free radicals as rust in our arteries.[10]

But how do you neutralize a radical oxide? The goal of eliminating free radicals has spurred the antioxidant craze, including blueberry smoothies, pomegranate juice and broccoli drinks. Luckily, antioxidants are readily available in all sorts of fruits and vegetables, and it is relatively easy to make them part of a healthy detox lifestyle. The simple message is to eat lots of fruits, vegetables, beans, nuts and whole grains. And blueberries, without question, make everyone's list as one of the most powerful antioxidants.

The general rule for antioxidant foods is to eat more darkly coloured berries and fruits (blackberries, raspberries, plums, prunes, etc.) as well as colourful beans (black beans, kidney beans, pinto beans, red beans), dark cruciferous vegetables (kale, broccoli) and whole grains. Garlic and tomatoes are also high in antioxidants—great news for lovers of Italian food like me. Avoid foods that are white and choose darker alternatives. Brown rice not white rice, whole grain bread not white bread, red grapes not green grapes, even red wine over white wine. (I have no trouble meeting my daily quota of that antioxidant!) Diets that are high in antioxidant foods help us to detox, protect us from cancer and improve cardiovascular health.

When healthy fruits and vegetables are hard to come by, melatonin, known mainly for aiding sleep patterns, is an important antioxidant supplement.[11] And all the health experts I spoke with recommended taking N-Acetyl Cysteine (NAC) supplements. NAC works by helping our body produce glutathione, a powerful antioxidant that neutralizes free radicals. Glutathione (GSH) is manufactured in our liver as a natural detoxing agent, and low levels can prevent us from detoxifying effectively. Glutathione, in sufficient quantities, even protects our cells from mercury damage.[12] The body's production of GSH depends on the intracellular availability of cysteine (an amino acid in the body that makes up the base of glutathione). Researchers at McGill University in Montreal have found that cysteine-rich supplements like NAC can help augment glutathione levels and can even enhance muscular performance during exercise.[13] The McGill researchers conducted an experiment with 20 subjects—10 who took cysteine supplements and 10 who took placebos. After the supplementation period, the two groups were tested for lymphocyte (cellular) glutathione levels and their peak muscular power. In both cases, the glutathione levels were significantly higher in the cysteine supplement group.[14]

The use of manufactured pharmaceuticals is a relatively recent phenomenon in medicine. People in all cultures have relied on natural remedies for thousands of years, and these are still at the core of traditional Eastern medicine. Many people my age will recall the cures for colds, coughs, aches and pains prescribed by our parents and grandparents: lemon and honey, hot toddies, gargling with salt water, drinking ginger tea.

The active ingredient in the ubiquitous aspirin (salicylic acid) occurs naturally in willow trees, codeine and morphine come from poppies and the pain relief ingredient capsaicine is found in chilis. About 40 to 60 percent of modern pharmaceuticals are derived from natural plants and/or synthesized plant derivatives. Plants are medicine.

So I'm always surprised when conventional doctors discredit herbal medicine as hocus-pocus. This is not to say that we need to believe the claims of all herbalists and homeopaths, but neither should we discount them outright.

Ayurveda is one of the oldest forms of herbal medicine, originating on the Indian subcontinent. Few scientific studies have been done on the efficacy of Ayurveda, but thousands of practitioners and many more patients swear by it. When I heard that Ayurveda was being used for detox treatments for the 9/11 first responders in New York, I wanted to know more.

I popped down to New York City to meet with two of the co-founders of Serving Those Who Serve (STWS), a volunteer organization dedicated to helping first responders to the 9/11 World Trade Center attack of 2001: José Mestre, the son of Cuban political prisoners who fled to Miami when he was a child, and Marshall Stackman, a jovial New Yorker with a solid build, greying ponytail and thick Brooklyn accent. We chatted for nearly two hours in a noisy, far too bright, fast-food restaurant on Columbus Avenue. We drank green tea.

José and Marshall recounted the tale of how STWS began after the horror and chaos of 9/11 and described the inch-thick dust that covered the cars in their Brooklyn neighbourhood. They asked me to imagine what was in that dust and smoke. "How many toxic chemicals?" they wondered aloud. "Hundreds? Thousands?" The Twin Towers attacks are considered to be among the most intense chemical exposure events in history.

And since we cannot deliberately expose people to high doses of toxic chemicals to see whether they get cancer or not, one of the techniques used by researchers is to study people who have been inadvertently exposed and see how they fare over time. Radiation victims in Chernobyl and mercury-poisoned residents of Iran and Japan are two groups that have been studied in this way. Much of what we know about the effects of chemical exposure on humans derives from studying large accidental settings like these.

Getting burned is not the main hazard connected with a modern fire. Somewhere between 50 and 80 percent of fire-related deaths are caused by smoke inhalation. When a house catches fire or when a car bursts into flames, one of the main concerns is the toxic chemicals released from the burning. The National Fire Protection Association reports that there are nearly 200,000 car fires on U.S. highways every year (that's more than 500 every day!), and close to a half a million buildings catch fire annually.[15]

Each time there's a fire in a car or modern home, a highly toxic mix of paint, vinyl, plastic, adhesive, foam, solvent and who knows what else ends up in the smoke. Look around your home and try to picture your television, computer, sofa, carpeting, cupboards, mattress, kids' toys, cleaning products, paint, wiring, insulation and roof shingles ablaze. As terrible as it would be from the standpoint of personal loss, imagine the toxic blend of fumes. This is what modern firefighters face on a daily basis. Volatile organic compounds (VOCs) are a major component of the toxic smoke and include well-known carcinogens such as benzene, toluene and naphthalene.[16]

If any group ought to be concerned about their chemical exposures, it is firefighters—especially the 9/11 first responders. The desperation and seriousness of the exposure resulting from this horrific event caused those affected to seek out any possible form of detoxification, and it turns out that saunas and herbal supplements have been among the most sought-after treatments.

The 9/11 attacks killed 4,000 people, and the long-term effects of the resulting toxic debris may be equally lethal. Public health officials admit that it is a huge and looming issue. One need only review the Centers for Disease Control (CDC) list of funded studies on 9/11 health impacts—literally, dozens. José claimed that an additional 2,500 people have died of related health issues in the 11 years since the 9/11 attack. The Ground Zero workers are being studied constantly, and an article published in the journal *Environmental Health Perspectives* found that firefighters with extended exposures at the World Trade Center site have significantly increased odds of showing respiratory and gastro-esophageal reflux disease (GERD) symptoms.[17] Studies of those present at the time of the attack have shown that within seven years of exposure, there was an increased risk of prostate cancer, thyroid cancer and myeloma.[18]

Cancer developing over the long term is one of the greatest scares, and firefighters know they have a "time bomb ticking" in their bodies. This was made very real for Marshall when he was told that 19 out of the 30 Trade Center rescue dogs died of cancer within just over a year of the attack. Marshall was also aware that cancers might not appear in humans for 10 years or more. For the majority of cancers the latency period (the length of time between an exposure and the onset of cancer) is 15 to 30 years.[19] A scary thought for anyone exposed to the toxic fallout at Ground Zero.

In the aftermath of the attack, several local organizations quickly surmised that there would be serious health problems due to the noxious exposures the first responders faced at Ground Zero. This was despite the official word at the time that the local air quality was not toxic, José recalled. Local churches and volunteers at Ground Zero knew this was a serious matter and immediately began setting up treatment centres for the debilitated rescue workers. Over time some of these facilities became more formalized, including Serving Those Who Serve, with their unique mission of delivering detox therapy through Ayurvedic healing. I confess my "flake-dar" went off at first, since I don't do yoga or follow Eastern religions much. Herbal treatments are at the core of Serving Those Who Serve, and they are combined with yoga, breathing and meditation to assist those who suffer from a wide range of health problems, including post-traumatic stress disorder.

I dug a bit deeper with José and Marshall. They described their early encounter with the physician who developed their herbal supplement protocol, renowned Ayurvedic physician Dr. Pankaj Naram. According to his website, "the original and authentic form of Ayurveda is based on one guiding principle: address the root cause of an ailment rather than its symptoms."[20] This seems to be pretty much in line with all forms of alternative medicine and once again highlights the contrast between naturopathic and ancient medicine on the one hand and Western medicine on the other, with its focus on isolating symptoms and prescribing cures, mostly in the form of pharmaceuticals.

Marshall and José recounted story after story of direct experiences with first responders whose health improved dramatically after taking the herbal treatments. José maintains that the herbs "bolster the immune system and enhance the body's natural detox processes." As with so many of these "alternative" programmes, it is still difficult to find scholarly articles with conclusive results that herbal detox programmes remove toxic chemicals.

I asked Marshall and José if anything had been written about their programme, and José directed me to a peer-reviewed study of 9/11 responders and locally affected workers who participated in the STWS Ayurvedic herbal supplement programme.[21] It is not a toxicology study showing the elimination of toxic compounds but a peer-reviewed survey of the participants' "post-treatment symptom impact"—meaning, Did the responders report feeling better after taking the herbal medicines?[22]

On a Likert scale of 1 to 6 for helpfulness (where 6 is the most helpful), the 50 patients participating in the study rated the herbal programme as being, on average, 50 percent more effective than conventional medicine.[23] José described what a challenge it is for these guys to take herbal supplements and undergo the Ayurveda treatments: "This is not a typical lifestyle choice [for them]." They are "tough guys," he said. "Staten Islanders, the hotbed of Republican support in New York City." Marshall added, "They eat poorly, smoke, drink, are often on prescription meds." Because firefighters historically see themselves as looking after others, they are reluctant to admit their own vulnerabilities and rarely stop to take care of themselves. However, since and STWS started doing its work, 1,300 of the 2,700 people who have participated in its herbal detox programmes are firefighters and STWS remains the only Ground Zero detox organization that is still up and running.

Canary in a Sweat Box

Armed with more information on complementary alternative medicine, body chemistry and liver function than I'd ever imagined existed, I felt ready to set off and do some personal detox experimentation. I was amazed when, in the testing we did for *Slow Death by Rubber Duck*, we could so easily manipulate our toxic chemical intake and actually measure the toxic chemicals that entered our bodies. But the big question for this book is will the reverse be true? Will I actually be able to measure the toxic chemicals *leaving* my body if I undergo specific detox treatments? My hunch was that it could be difficult to pull off and that "tox-in" testing would be more predictable than the "tox-out" version. But challenges of this nature had never stopped us before! So on we went.

Many of the wide range of detox strategies I've reviewed are not easily measurable. Most off-the-shelf cleanses are not specific enough for their effects to be measured, nor is there any evidence to suggest that they really work. Dietary changes are too subtle and incremental. Herbal detox therapies don't allow for observing the excretion of toxins either. And there is the difficulty of controlling inadvertent exposure to chemicals, a problem that Rick and I face regardless of how careful we are. One brush with a heavily perfumed individual could easily result in perceptible increases in phthalate levels that would skew the results of any experiment.

Were there any popular detox treatments I could try out that would allow for accurate measurement of toxic chemicals leaving my body? I was beginning to wonder. But then the idea of saunas came up. Taking a sauna would be one reasonably accessible, proven therapy for accelerating detoxification that would allow for testing. And how unpleasant could that experiment be?

First things first. I needed to get my hands on an actual sauna unit. I contacted Rodney Palmer, an infrared sauna manufacturer, whom I'd met at a health show. In addition to kindly lending me an infrared sauna for the experiments, Rodney provided me with a wealth of information about sauna detox therapy. What's more, like so many people immersed in alternative health, Rodney had his own story to tell.

I invited Rodney over to my house, and we sat on the back deck, drinking pomegranate juice mixed with filtered, carbonated tap water.

Rodney began to recount his tale. He took me back to 2003 and the SARS (Severe Acute Respiratory Syndrome) epidemic. Rodney did not have SARS, but during the outbreak he was living in Beijing with his young family, covering the story as a Canadian journalist for a major television network. There was a Canada–China connection after the SARS outbreak originated in China when a woman travelling from Hong Kong to Toronto brought the disease to North America. She died shortly after arriving in Toronto, as did her son, but not before infecting numerous healthcare workers and making Toronto the most deadly SARS destination outside of Asia.

Rodney was reporting initially on China's refusal to acknowledge the presence of the highly infectious killer disease and later on the hundreds of deaths and the global response to the SARS epidemic.

But SARS wasn't the only airborne threat to health: China has some of the world's worst air pollution. "You couldn't see across the street. You could tell there was a building there, but you couldn't tell how tall it was or how many windows were in it; it was really, really bad." Looking back, Rodney wondered if the health problems his family would ultimately experience were the result of toxic exposure to the gross pollution in China or perhaps to the lead plumbing in Israel, where he'd been assigned previously. Or were their health issues the result of a combination of many exposures over the years? This question of what causes health problems haunts every person suffering from environmental contamination or chemical sensitivity. Rarely is there a definitive answer. The salient matter for Rodney, though, was that his wife's and son's symptoms first appeared when they were in China.

The decline in their physical well-being began at roughly the same time, with his wife's health deteriorating most dramatically.

"She was losing weight rapidly, and there was this scaly stuff on her fingers," Rodney recalls. But his son was affected too. "My son wasn't growing properly. He had problems with his leg joints, and he couldn't even hold his own weight." Rodney described how they were almost going crazy trying to figure out what was happening. His wife had lost 40 pounds, could no longer work and was manic. They were both very worried for their son.

At this point I felt as though I could have finished the story for him—it was so reminiscent of Peter Sullivan's experience that it was eerie. Sure enough, Rodney described how his wife came across Sherry Rogers's book *Detoxify or Die*, which led her to have the family tested for toxic contaminants. It turns out they all had hugely elevated levels of lead in their systems. Lead is a neurotoxin that causes a variety of mental impairments, and it is also associated with dozens of other serious afflictions, including gastrointestinal problems, kidney disease, anemia and even death. The doctors they consulted in North America suspected that lead had been replacing calcium in his son's growing bones, causing them to soften—hence his difficulty with walking. They attributed his wife's weight loss, her anemia and her frantic mental state to elevated lead levels. Rodney had no symptoms at all despite his elevated lead levels, reinforcing the point that we all respond very differently to toxins, some people having much higher tolerances.

Sauna to the Rescue

Rodney's family was introduced to Dr. William Rae, a cardiovascular surgeon and considered to be one of the fathers of environmental medicine. When he learned that Rodney's wife had elevated levels of lead, he administered an intensive course of sauna therapy. Within three weeks she was feeling better—not perfect "but no longer crazy," Rodney said, smiling. He assumed that a combination of getting out of the toxic environment in China and undergoing sauna therapy must have helped her.

Without easy access to a sauna, Rodney purchased an infrared sauna kit to install in their home, but he was horrified when he turned it on and found that it smelled of airplane glue. After extensive research he discovered that most infrared sauna kits were poorly constructed and contained toxic glues and electrical components with high EMF emissions.

Some people may have given up at that point, but not Rodney. He took things to the next level and decided to design and build his own infrared sauna—one without any glues (only screws as fasteners) and with wood and electrical components that he sourced and tested personally. And that's how his company, SaunaRay, was born. Before too long, environmental doctors were asking him about his "clean" infrared sauna, and he now sells his sauna kits around the world.

Rodney was so convinced that sauna therapy was one of the most effective ways of removing toxins from our bodies that he took issue with one of the claims Rick and I were making in *Slow Death*: that if people stop using toxic products they will naturally eliminate them and our levels will return to their original state. Rodney was certain that toxic chemicals did, in fact, get "stuck" in our bodies, and as proof he pointed to levels of toxins in sweat being sometimes much higher than levels seen in blood or urine. A research paper by Dr. Genuis, the detox expert I've mentioned several times already, has described the same concept but in technical terms: "[Blood] serum levels of various xenobiotics [chemicals that are foreign to living things] do not necessarily reflect the total body burden of such compounds because accrued toxicants may store in tissues, and serum levels may belie actual toxicant status."[24]

Dr. Genuis also helped me design the sauna experiment and was invaluable when it came to teaching me the basics of sauna detox and how to create a protocol for testing and measuring my sweat.

Much of the detox world is filled with people who have compelling personal stories. These are important, often heart-rending and life-changing experiences for the individuals who live through them, but at the end of the day they still constitute just one person's experience, and it's difficult to apply such experiences to a larger population. A clinical doctor, on the other hand, treats hundreds of patients and begins to see trends in both symptoms and treatments. Clinicians doing detox medicine are at the forefront of the health field and deserve tremendous respect for their pioneering spirit.

Now back to the experiment. The sauna protocol I developed based on Dr. Genuis's advice included running, followed by sauna therapy and the ingestion of vitamins, psyllium husk fibre and electrolytes. The psyllium husk helps to absorb and bind the toxins, aiding in their removal. Electrolytes, such as sodium and magnesium, are critical to our health and can be depleted to dangerously low levels with vigorous sweating. Gatorade, for example, was invented by researchers at the University of Florida as a way of rehydrating and replacing lost electrolytes for athletes.

Rick and I are often asked whether an individual can get themselves tested for toxic chemicals. Theoretically, the answer is yes. Practically speaking, however, the situation is much more complex. First, the tests are very expensive (thousands of dollars) and not undertaken routinely by doctors or labs. This means that approaches to testing and analyses vary. Second, large research labs do not seem to be well equipped to handle small numbers of samples from individuals or even from individual doctors' practices. The best way to have tests done on a broad spectrum of chemicals in your body is to find a doctor who specializes in environmental or integrative medicine.

That said, it's very difficult to find a research lab that will test levels of rare synthetic toxic chemicals in sweat, partly because it's much more challenging to test sweat than blood or urine. I undertook nearly two months of lengthy and detailed discussions about sample protocols and research parameters with several labs before I found one that was willing to perform the sweat analysis for us. At $900 a pop for some of the samples and the complicated conversations with the labs about the chemistry behind the procedures, this kind of testing is far from routine or accessible to the general public.[25]

The Big Sweat

The closet-like sauna was installed in my basement, its small, sliding-glass window facing the furnace. It was cramped but just the right size for the small utility room. With my sample jars and vials sorted out (no easy task), I was ready to start the first sauna sweat collection. The protocol called for two one-hour sweats per day over a three-week period, with weekly collections. This was modified to a maximum of one sweat per day over a five-week period, for reasons described below. I was also collecting urine to compare toxic levels before and after sauna for the duration of the sauna treatment. And on top of that, the plan was to collect my blood to compare levels in my blood versus levels in my urine and in my sweat, attempting to emulate studies that Dr. Genuis had conducted.[26]

Before entering the sauna, I drank a large glass of water (about 500 mL) with a scoop of electrolyte powder added to it. Then I collected a 100-mL urine sample and weighed in at 198 pounds. I'd completed a 20-minute run before getting into the small, wooden cubicle, to get my blood flowing, and within minutes of stepping in, plenty of perspiration was flowing out of my body. The temperature was 40 degrees Celsius, and after five minutes or so, beads of sweat were dripping off my forehead and onto the floor. My run before the sauna had helped increase my internal body temperature, and that was making small, distinct droplets appear on my forearms—the first sign of a good sweat.

It turns out that medical researchers have been collecting sweat from forearms since as far back as 1934 because of interest in the relationship between sweat-gland activity and blood flow in the forearms. Research I have read indicates that sweating happens first, and then the blood vessels dilate—never vice versa—and when the blood vessels dilate, blood flow increases. So not only does the heat from a sauna provide a mechanism for certain synthetic chemicals to be excreted through sweat, but the increased flow of blood helps our livers process and detoxify a greater volume of blood. And we know that the liver is the body's primary organ for breaking down and eliminating toxic chemicals. Sauna therapy therefore enhances detoxification in two ways: by inducing sweating and by increasing blood flow.

After 10 minutes at 46 degrees Celsius, the sweat was really flowing. I was pleased with the performance (based on my sweat volume), and I was feeling great.

Never having collected my sweat in a glass jar before, I wasn't entirely sure how I would go about it. But I just scraped the edge of the jar along my skin, hoping that the sweat beads would find their way into the container. Before I knew it, I'd managed to collect a discernible volume. Seeing a 250-mL (8-ounce) jar nearly half full of sweat was somewhat bizarre. It made me wonder how much I was actually sweating out in total. If I'd collected nearly 100 mL (3 ounces) of liquid, and assuming I captured around 5 to 10 percent of my output, I must have sweated out a total of between one and two litres. But that seemed extreme.

I left the sauna, weighed myself, had a shower and drank more water with electrolytes added. According to my bathroom scale, I now weighed 195 pounds. My one-hour sauna had sucked three pounds of sweat out of me. Was this possible? Using my estimate of between one and two litres of sweat collected, I did the math, and it was indeed possible.[27]

After my hour of sauna therapy on day one, I did another hour on day two. I didn't feel so well coming out of the cubicle on either day, I was drowsy most of day one despite consuming large amounts of water, electrolyte supplements and psyllium husk powder. Day two was even worse, as I was basically immobilized, unable to function for several hours. I started to wonder whether this was some sort of healing crisis and if saunas were even a good idea.

Day three, and it was time for a blood sample. I'm not a big fan of giving blood at the best of times, and having dreamt the night before that the blood sampling didn't go smoothly, I wasn't looking forward to the procedure. But when the nurse arrived, there was no question we were going ahead and he began describing the bloodletting procedure in colourful detail, something I was not prepared for. I sat down, the nurse tightened the tourniquet around my arm, and then he began poking at veins on the top of my hand. Poke, poke, poke and still no blood. After several more attempts I started feeling very queasy. "I'm not sure what's going on, but your blood is very slow," the nurse observed. I could feel a nauseating light-headedness start to wash over me. I knew what was happening and told the nurse I was about to pass out. I went with the flow, or lack thereof, and laid my head back on the sofa as I blacked out completely.

I was conscious again when the 911 gang arrived: first the firemen, followed by the paramedics. I was also sweating profusely. Just what I needed. More sweat. The battery of tests began. Given that I am now a post-50-year-old male, they were anxious to see how my heart was doing. First they checked my blood pressure: 127 over 82. "Fine," I was told. They did a full 12-lead electrocardiogram while at the same time testing my blood for diabetes. "Not diabetic," one of the paramedics said. My dog sensed something was wrong and would not leave my side, lying against me on the sofa. "Heart looks good, heartbeat normal, beating strongly," said the guy reading a cardiogram printout on the tiny graph paper spewing out of the portable machine on my family room floor. I thanked them and declined a ride to the hospital.

I discussed with Rodney the fact that I was feeling rather poorly after my two long sauna sessions and mentioned the fainting incident and the 911 call. "I wondered if perhaps I was overdoing it," I asked, thinking of the one-hour saunas at up to 49 degrees Celsius. Rodney's reaction was quick and showed that he was worried: "Yes, of course, you're overdoing it." He told me to keep the temperature lower, open the sliding-glass window in the cubicle for fresh air, pull back to 30 minutes and work my way up to an hour. He also told me to make sure I was always well hydrated and to take adequate electrolyte replacements and vitamins.

The irony of the experiment was not lost on me. In *Slow Death*, Rick and I had exposed ourselves to a bunch of nasty chemicals, and truth be told, we did not really feel too much the worse for wear after that. Essentially, we had to live with some nasty smells for a while, and I may have had some minor mood swings resulting from my near-mercury poisoning. Yet it felt as if the detox experiments were killing me! I could barely function. Rodney warned me that the rapid rate of toxins being released via my aggressive sweating routine could result in temporary discomfort, but I have a hunch it was mainly dehydration.

The experience reminded me that despite the incredible resilience of our bodies, we live within remarkably narrow tolerance ranges related to body temperature, hydration, salt levels, blood pressure and many other parameters. It's easier than we think to throw ourselves out of balance—beyond the defined ranges that sustain our lives, and this can pose serious health risks. Detox programmes need to be undertaken with caution and according to the advice of a trusted health practitioner.

Well, What *Was* in My Sweat?

Undertaking these analyses was stressful. Rick and I spent a huge amount of time and money trying to figure out every last detail, but because we were doing things that had never been done before, we had very little idea of how they would work out. So to keep things manageable in the blood-and-sweat sauna experiment, we decided to focus on two of the chemicals that we'd researched extensively in *Slow Death*: phthalates and bisphenol A (BPA). Phthalates, as we've mentioned, are widespread in personal-care products and in air fresheners used as artificial scent, and BPA is an ingredient in the plastic resin of thousands of products from the visors of hockey helmets to the lining of food cans.

BPA (see Figure 8) in both my sweat and urine was measured at 1.7 ng/mL in the first sample, which is about the mean level found in American men.[28] Over the course of the first three weeks, my BPA measurements remained fairly stable in both sweat and urine, decreasing slightly over that time period. My daily routine must have been quite consistent. But the levels found in week four were shocking. My BPA measurement of 4 ng/mL of urine put me in the 90th percentile of Americans, and although there are no statistics on BPA in sweat, the levels found in my week-four sweat had skyrocketed to 14 ng/mL. A flaw in our research became obvious. I hadn't kept track of my diet or bathing routines during the experiment. Generally speaking, I avoid BPA and phthalate exposure to the extent that I can control such exposure, with one weakness: I love to cook with imported, canned San Marzano tomatoes, and those cans almost certainly contain BPA in the lining. I'll bet I had a large helping of tomato pasta made with those canned tomatoes the evening before the BPA spike.

Figure 8. BPA levels in Bruce's sweat and urine over five-week period (ng/mL)

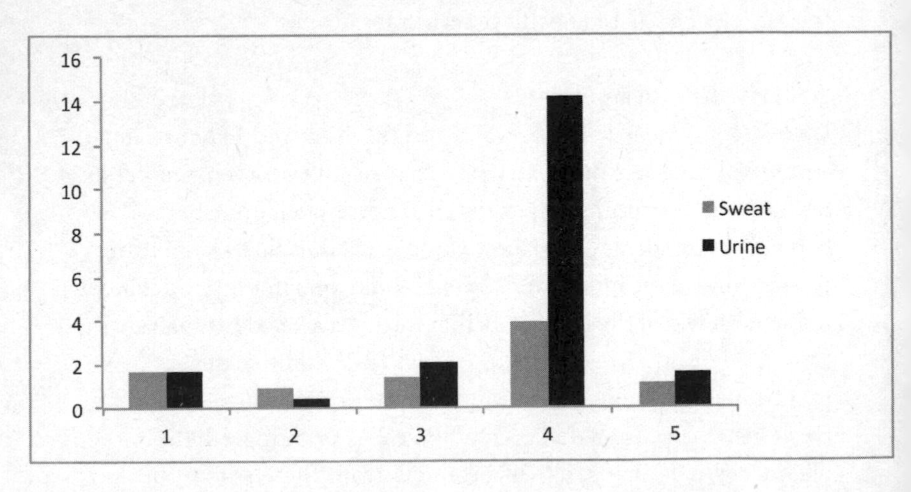

The results of my phthalate-level testing resembled those of the BPA tests, in that there was no consistency across the five-week study period. This is not entirely surprising given that these samples are a snapshot in time. MEP and MBP, the urinary metabolites of diethyl phthalate (DEP) and dibutyl phthalate (DBP), respectively, are widespread in personal-care products and cosmetics. Though I've been steadfast about eliminating scented soaps since the last book, Figures 9 and 10 show that these chemicals were still very much present in my body. My urine levels of MBP were consistent with the American mean of 33.4 ng/mL, disregarding the spike in the last week, which put me well above the 95th percentile (did I grab the wrong shampoo one day?). And my MEP levels were well below the 1017.5 ng/mL mean, even with the significant week-three spike (see Figure 9).[29] I attribute those low MEP levels to my successful strategy of scented-soap avoidance.

Figure 9. Phthalate metabolites (MEP and MBP) levels in Bruce's urine over five-week period (ng/mL)

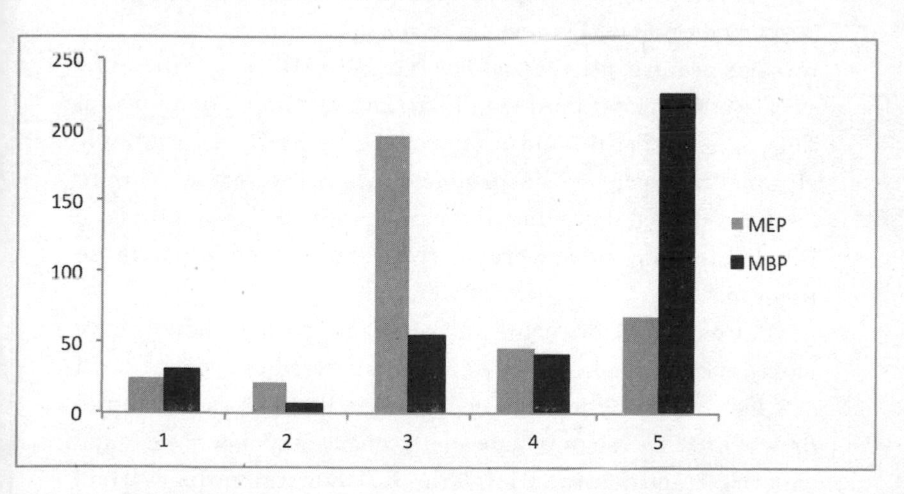

Figure 10. Phthalate metabolites (MEP and MBP) levels in Bruce's sweat over five-week period (ng/mL)

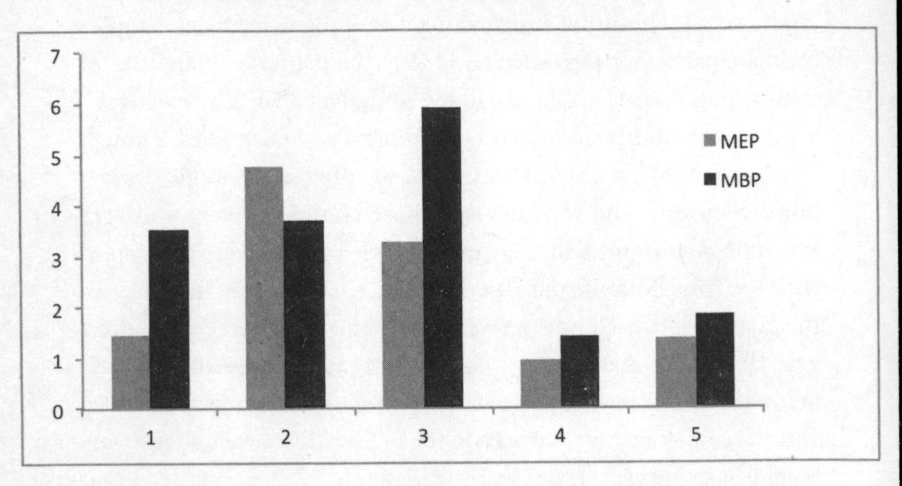

My sweat and urine levels for BPA and phthalates both followed similar curves, suggesting that testing for either urine or sweat provides an accurate indication of relative levels over a particular period of time. However, the most important information revealed by the testing is that levels of both MEP and MBP in my sweat were considerably lower than the levels found in my urine. This implies that the kidneys provide the preferred route for eliminating phthalate by-products, whereas sweat appears to be the preferred detoxification pathway for BPA, given that the BPA levels in my sweat were nearly five times higher than those in my urine.

Dr. Genuis's blood, urine and sweat monitoring studies[30] have looked at levels of selected toxic chemicals in sauna-induced sweat as a biomonitoring tool for body burden levels. It is difficult to draw a direct correlation between the quantity of chemicals found in our sweat and the overall levels in our body over time. Inadvertent daily exposures may result in levels that spike off the charts, as my test results showed. Measuring toxins in sweat confirms that we have specific chemicals in our bodies and that we can help rid ourselves of them, some more than others, by sweating them out. And that is a good thing.

Despite the little research hiccups, my over-enthusiasm in cranking up the sauna heat and staying in too long too soon (to say nothing of the fainting episode), saunas (done the right way) make me feel great. I leave a sauna feeling clean, relaxed and serene. And my skin always feels amazing. After my intense sauna routine, people I barely knew were commenting on my skin. It was kind of weird. Out of the blue people would say (unprompted), "Wow, your skin looks so fresh and healthy." Given the outrageous amount of money people spend on skin and face lotions and creams, I suspect that investing in an infrared sauna and using it twice a week might actually save money, produce better results over the long term and also be safer and healthier than using personal-care products laced with harmful synthetic chemicals.

Feeling the Detox Boost

With my successful chelation results from the previous chapter in hand, along with this evidence that sweating is a great way to rid our bodies of synthetic toxins, and having consumed more water in the previous 6 months than in the 50 years that preceded them (or so it felt!), I was starting to understand the concept of "toxout" and could see that in my case at least, chelation and sweating in a sauna were generally effective methods of getting synthetic chemicals out of the body. True, the chelation also took out sodium, calcium, magnesium and sulphur—elements that the body needs. And the sauna routine had its ups and downs. However, overall, they were extremely useful.

Detox methods like chelation and sauna-induced sweating could help the meaningful percentage of the population who suffer from chemical sensitivity, as well as various allergies and other reactive sensitivities. The key word is "sensitivity." We don't hear people being described as having "migraine sensitivity" or "cancer sensitivity" but these are not dissimilar concepts. Chelation is an important treatment for people with elevated heavy metal levels or heavy metal sensitivity. Sauna-induced sweating can complement detox therapies as part of a specified treatment routine or as a general preventive strategy for anyone concerned about exposure to toxic chemicals.

And now a final word about sweat.

Humans wouldn't spend US$18 billion per year on deodorants and antiperspirants if we liked sweat. But is perspiration getting a bad rap? Rather than thinking of sweat as disgusting, in terms of prevention of cancer and other disease, we should be thinking that *not* sweating is disgusting. Sweating is one of our body's most natural processes. It helps regulate our fluid and electrolyte levels and eliminate toxins by taking waste from our blood and expelling it. Think of our pores, as I witnessed in the sauna, as little springs releasing droplets of water, salt and unwanted toxins. And we do more than sweat through our skin. Our bodies also release oils through sebaceous glands, and these oils carry away fat-soluble toxins. Imagine if all those glands and pores were plugged with makeup, creams or even worse, aluminum-based antiperspirants designed to shut them down. I liken the plugging-up process to spreading plaster wall filler over nail holes. It leaves no place for anything to come out. It stops up our sweat and holds in all the toxins and oils in our bodies, and this leads to aggravated skin, clogged pores, acne and other dermatological problems—not to mention preventing our body from using the full potential of this critical detox pathway.

Detoxing is a constant process for all humans. It happens to some extent whether we do anything about it or not. Every minute of every day, we are expelling natural and synthetic toxins. So the beauty of detox therapies is that they are designed to improve and enhance our existing detox systems and pathways. All they really do is give our bodies a little extra help so that every day we flush out more of the nasties than we would naturally.

The critical piece of that statement is "every day." This means eating organic fruits and vegetables; cutting back on animal fats;

drinking lots of fresh, filtered water; taking relevant supplements and exercising *every day*. And if you can fit a sauna into your routine, all the better! Detoxing is not about one-off diets, cleanses or foot bath fads; it's a lifestyle we need to subscribe to. So save some money and avoid the cleanse kits.

We need to push the reset button on our toxic lifestyle to keep our toxout fitness levels high.

FIVE: A STATIONARY ROAD TRIP

~ Rick breathes deeply ~

Advertising is based on one thing: happiness. And do you know what happiness is? Happiness is the smell of a new car. It's freedom from fear. It's a billboard on the side of a road that screams with reassurance that whatever you're doing is OK.

—DON DRAPER, *Mad Men*

DON'T GET ME WRONG: Jeff Gearhart is a lovely man.

I just never want to see him again.

This is what tends to happen when you spend a day sitting beside somebody in a car parked in a hot room, with the windows rolled up, inhaling toxic fumes.

People start to wear on each other.

Thankfully, the stuffy vehicle of which I speak was not the fire-engine-red Fiat 500 that I rented at the Detroit Metro Airport and drove to Gearhart's Ann Arbor, Michigan, office. Though it was tonnes of fun booting along the I-96 on a crisp and sunny November day—and certainly a sportier ride than what I normally have in my VW station wagon—lingering a further eight hours in the little car would have left my six-foot-six frame twisted up like a pretzel.

No, the car Gearhart and I spent so much time in was pretty much the polar opposite of a Fiat. Standing at over 9 feet high and 17 feet long and weighing in at a whopping 7,100 pounds, the

Chevy Tahoe LT is a real All-American vehicle. Or at least two-thirds "All-American" according to the manufacturing information.[1] Sixty-seven percent of those steel, plastic, aluminum, rubber and glass parts were made in the U.S.A. and assembled in Arlington, Texas, before being shipped to General Motors' central distribution centre in Lansing—just an hour from where I would spend my fateful day in its confines.

The specific Tahoe in question was jet black and brand spanking new. It showed just 274 miles on the odometer, and the tag on the door told us it was built about a month before Jeff rented it. All in all it had everything we needed for our strange day: a huge interior we could stretch out in and, most important, an overpowering, nose-tingling "new car smell."

Ahhh . . . the "new car smell."

So intoxicating, yet so fleeting. Is your aging beater getting you down? No problem! You now have the choice of a veritable smorgasbord of "New Car Smell"–scented air fresheners for that quick vehicular pick-me-up.[2] And if you can't get your car to smell the way you like it, you can always smell like your car: The new "Bentley for Men" cologne is said to evoke its fancy car namesake's aroma of "fine woody notes and exquisite leather."[3]

I can assure you that no such crass manipulations were necessary that day in Ann Arbor. The Tahoe's acrid interior didn't need the boost of air fresheners or colognes: It stank to high heaven all on its own. Remember back in chemistry class . . . those mixtures you had to concoct in the laboratory beakers? Just think of the Tahoe that way: as an experimental vessel. Add a couple of human guinea pigs to its noxious contents, apply a bit of heat *et voilà!* A first-of-its-kind experiment that would make Mr. Wilkinson—my high school science teacher—proud.

Did we follow the "scientific method"? Well, sort of. We had a hypothesis: that the new car smell would result in measurable increases in chemicals in our body. We had an experimental design. And very important, if anyone cares to replicate our results, we've provided enough detail that they can. Would wallowing in the SUV's pungent embrace fill our bodies with pollutants? We were about to find out.

Space Station

The cool factor of the Ecology Center's stylish walk-up office is seriously undermined by the shop selling "discount ugly Christmas sweaters" down below. Fortunately, Jeff Gearhart, the center's research director, was not wearing one when he greeted me at the door. He was, however, wearing a backpack stuffed full of delicious food to sustain us throughout the day. We wasted no time in piling into his recently rented Tahoe and hitting the road for the 40-minute drive to "The Space Station" self-storage facility in Petersburg, Michigan, the closest place that Jeff could find that had a room big enough for our experiment.

"So what are you up to again?" asked Leanne, the proprietor, who took our rental payment for a day's use of her large, heated garage.

"We're measuring air quality in cars," Jeff offered. "We want to mimic the cabin temperature on a hot, sunny day, so don't be surprised if you come into the garage and it's a little bit warm."

"Riiighht," she said. Looking us up and down, eyebrow raised, as she rang Jeff's credit card through. Just a hunch, but I wouldn't imagine she got a lot of customers looking to stay and bask in her storage lockers for extended periods of time.

Jeff and I drove the Tahoe into the middle of Leanne's garage, then cranked the thermostat on the wall to 86 degrees Fahrenheit (a common temperature on a summer's day in the Ann Arbor area). We rolled up the windows, turned off the car, tuned the radio to NPR and sat there for the next eight hours, breathing in the pungent off-gassing from the vehicle's newly minted upholstery.

I've heard the most powerful memories are triggered by what you smell. And I'll admit that for me, the "new car smell" took me right back to wondrous family holidays in Bar Harbor, Maine: the sea air, sunlight reflecting off the pink granite mountains of

Acadia National Park and watching the world go by through the window of my dad's new 1972 baby-blue Dodge Dart Swinger. These were some of the sepia-tone images conjured up by the Tahoe's fragrance.

On a molecular level, however, the "new car smell" is considerably less magical: It's nothing more than nasty volatile organic compounds (VOCs) (like formaldehyde, toluene and xylenes) and other chemical goodies evaporating from the adhesives, sealants and plastic bits in the vehicle's interior. "Enjoying the smell of a new car is like 'glue-sniffing,'" read a pithy article on the subject by the eminent environmental journalist Charles Clover in the *Telegraph*.[4] Being a biologist by training, when I hear the word "formaldehyde," I also think of the jars of pickled dead animals that lined the walls of my university classrooms.

The structure of our experiment couldn't have been simpler. Before we left Jeff's office, we each gave a urine sample. And when we climbed out of the Tahoe in the early evening, we each gave a second sample. We then sent the four little glass jars off to EAG Life Sciences, an expert lab in Maryland Heights, Missouri, to measure whether the eight hours of breathing in the "new car smell" appreciably increased the levels of certain VOCs in our body.

If any of you have ever spent eight straight hours in a hot car with another person, you'll sympathize that the conversation can pretty easily go off the figurative road, but Jeff and I did manage to chat at some length about his work at the Ecology Center and the reason why we were in the car in the first place. The Ecology Center's work on cars and chemicals started at a grassroots level in the communities impacted by manufacturing plants in southeast Michigan. The center realized it could leverage broader public support by figuring in the "iconic product" of this manufacturing—the car that almost everyone was driving—and tie it to their issues. So in the late 1990s they started doing report cards that examined and graded the car manufacturing process. These reports were quite broad in scope, assessing environmental factors and corporate management to determine which companies were doing better than others. Jeff told me that after a few years of preparing these report cards, they began to realize they needed to figure out a way to make things more real for people. So the center developed a new method of materials testing that allowed for the benchmarking of every vehicle on the market.

The first *Healthy Stuff* new vehicle consumer guide was released in 2007—to so much interest that the host website crashed. The guide reviewed over two hundred of the most popular car models from that year, testing each of them for the toxic chemicals that off-gas from the interior parts: the steering wheel, dashboard, armrests and seats. The *Healthy Stuff* vehicle guide focuses on chemicals such as flame retardants, plasticizers like phthalates, heavy metals and VOCs. "Automobiles function as chemical reactors, creating one of the most hazardous environments we spend time in," Jeff said in one of the center's latest press releases. To sample the vehicle interiors for these chemicals, Jeff and his team developed a method using a portable X-ray fluorescence (XRF) device, an impressive machine that slightly resembles a chunky phaser from *Star Trek*'s USS *Enterprise*. When you hold the XRF up to a piece of material, it identifies the elemental composition of the thing it's aiming at in less than 60 seconds. Each car is then ranked according to the results of this material analysis, with the worst cars receiving high scores due to their high chemical content.

Over the last six years the Ecology Center has released four consumer guides, testing over two hundred models each year that a guide was produced. To date they have about one thousand cars in their database, and Jeff estimates they have about 90 percent of the vehicle models in North America from 2006 and on. Every year, Jeff and his team conduct their exhaustive measurements, and every year an alarming cocktail of toxic chemicals is found in the interior of many cars. But there is some good news: Overall ratings are improving and the best vehicles have eliminated some hazardous chemicals. In 2012, for instance, 17 percent of new vehicles tested by the center had PVC-free interiors, and 60 percent were produced without brominated flame retardants.[5] Jeff told me that car companies like Ford are now investing significant dollars to clean up their manufacturing act, and Honda was so delighted that its Civic was named the Ecology Center's 2012 "Best Pick" in recognition of the company's work to make its vehicle cabins as non-toxic as possible that it now regularly trumpets this accomplishment on social media. It's also considering printing up stickers with the Ecology Center's logo to be displayed at its dealerships. To quote the old song, the Ecology Center "ant" has moved the car industry "rubber tree plant" in a very significant way.

None of this good news applies to the Tahoe, unfortunately. In its most recent report the center classifies the car as of "medium concern" and found measurable levels of bromine, chlorine and other nasty chemicals in its interior. And it's not only toxicity that figures among the Tahoe's environmental misdemeanours: This is a vehicle that seems determined to poke Mother Nature in the eye. From its introduction in 1996 through to 2002, the Tahoe's popularity increased—reaching a peak in 2002 of over 209,000 vehicles sold. But since that time the Tahoe has fallen victim to increasing demands for improved fuel economy. With a current average fuel efficiency of 17 miles per gallon (7.2 km per litre) as compared to the 54.5 miles per gallon (23.2 km per litre) that the United States and Canada are aiming to achieve by 2025, the Tahoe is increasingly out of step with the automotive zeitgeist. As a consequence, 2012 sales of the SUV were less than a third of what they were 10 years prior.

I read all this and much, much more about the Tahoe as I sat beside Jeff during the eight hours we were ensconced in its capacious interior. Though some University of Michigan graduate students who work with Jeff busied themselves taking air samples from the car's interior, they came and went in a flash, leaving Jeff and me alone, staring at the wall of the storage depot. I can report that the Tahoe has side-view windows that move in and out electronically. Playing with them was good for 15 minutes or so. I thoroughly explored the website of Ann Arbor's famous Zingerman's Deli and daydreamed about delicious take-out sandwiches. And with Christmas coming up, I had plenty of time to concoct a more imaginative shopping list than I had in years. No socks for my hard-to-buy-for diabetic, 95-year-old aunt Pat this year. No, sir! This holiday season I located the world's most delicious basket of handmade, sugar-free chocolate: a Yuletide home run for sure. Thank you, Chevy Tahoe.

Regardless of the quality downtime it afforded us, by hour eight Jeff and I were hanging on every minute as the dashboard clock clicked down to the end of the experiment. We wheeled out of the garage like bats out of hell, racing through the darkness back to Ann Arbor, courting a speeding ticket, so anxious were we to be free of our vehicular incarceration. The new car smell had gotten stale.

Sniff Test

That evening Jeff, his wife and I retired to a local bar to celebrate the end of our stationary road trip and speculated on the outcome of our weird day. As with many of our efforts in this book and *Slow Death by Rubber Duck*, nobody had done this sort of thing before, and we literally had no idea what to expect. Yes, there have been a lot of studies of indoor air quality—and the reason, in this couch potato age, is likely obvious: The average resident of an industrialized country spends more than 90 percent of his or her life in enclosed spaces. To be even more precise, Americans have been found to spend almost 87 percent of their time indoors and an incredible 5.5 percent of their lives in an enclosed vehicle. For those of you who aren't fans of math, this leaves only about 8 percent of our lives for skiing, raking leaves in the autumn, sunbathing on docks, walking kids to school, golfing, enjoying picnics and all other outdoor pursuits combined (Figure 11).[6]

Figure 11. Where Americans spend their lives

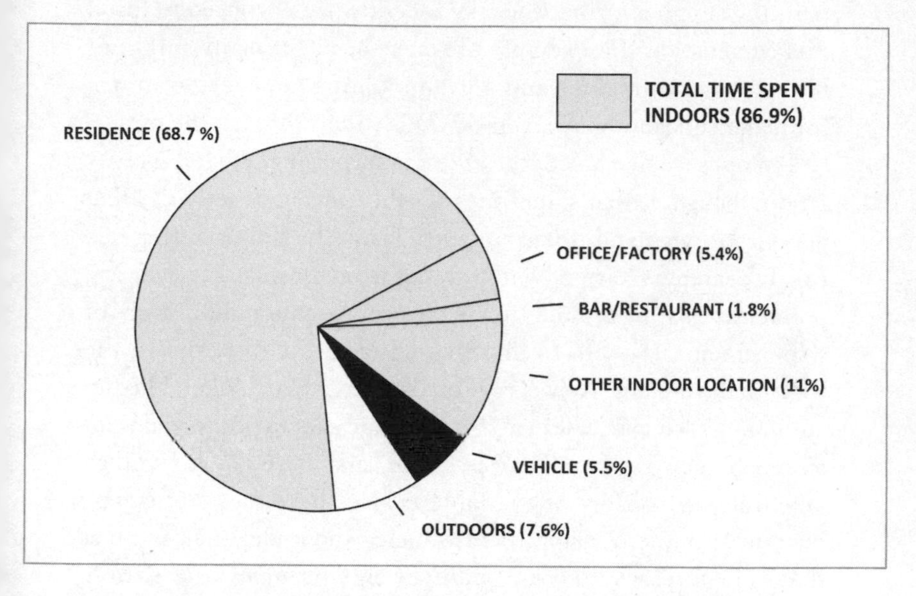

Adapted from N. Klepeis, W. Nelson, J. Robinson, A. Tsang, P. Switzer, J. Behar, S. Hern and W. Engelman, "The National Human Activity Pattern Survey (NHAPS): A Resource for Assessing Exposure to Environmental Pollutants," *Journal of Exposure Analysis and Environmental Epidemiology* 11 (2001): 231–52.

Scientists have increasingly turned their attention to indoor air pollution because, quite simply, this is where much of the world is spending their lives.

Indoor air pollutants are an incredibly varied lot, and virtually every American (and, it would be reasonable to assume, residents of other industrialized nations) have measurable levels of VOCs and other pollutants in their bodies.[7] Some indoor activities expose you to more pollutants than others. Not surprisingly, the presence or absence of tobacco smoke is a huge determinant of indoor air quality. So is whether or not your home has an attached garage (because of its inevitable and smelly assortment of jerry cans, lawn mowers, broken-down cars and sundry abandoned painting supplies). And if you've recently been exposed to fossil fuel combustion (for example, if you've been sitting in traffic) or have worked with glues and solvents (doing home renovations, for instance), your body levels of VOCs will be through the roof.

I'm not sure that any of the above is surprising. What is a revelation, though, is that sometimes it's the innocuous activities that also contribute to our toxic chemical load: like simply sitting in a car. Researchers have now found that we are exposed to over 275 pollutants floating around inside the average automobile. More of these chemicals seem to be released from car interiors in the warmth of summer. New cars less than three years old and luxury automobiles (maybe because if you've paid a lot for a car, you want as much "new car smell" as possible) also have above-average smelliness. A variety of available studies have measured toxic chemicals in the interior air of vehicles and underlined the fact that in many cases their air quality is significantly worse than is considered safe in indoor environments. However, it hasn't been at all clear that these ambient levels would result in measurable increases in the human body.[8] This was the important issue Jeff and I wanted to investigate because it goes to the heart of what this book is all about: The first step on the way to *toxout* is knowing what chemicals to *avoid* in the first place.

Of all the chemicals we could have looked for (because, as just mentioned, there are hundreds of individual VOCs and other potential pollutants in indoor air) we zeroed in on specific chemicals that are known to pose health concerns.[9] We looked for the metabolites, or breakdown products, of hexane (2,5-hexanedione); heptane (2,5-heptanedione); benzene (phenyl mercapturic acid); toluene (benzyl mercapturic acid); and xylenes (methylbenzyl mercapturic acid). Hexane is one of the most common VOCs in many glues, and prolonged exposure to it can lead to neurological damage. Heptane is common in some paints and coatings and certain kinds of rubber cement. Benzene is a well-known human carcinogen, and exposure to toluene and xylene can cause changes in the central nervous system and other damaging neurological effects.

The results of our experiment were stunning (see Figure 12). For all the chemicals in question (including benzene, not shown here), both Jeff and I saw significant increases in our bodies after eight hours of breathing the off-gassing. Look at the trends: In all cases our VOC levels rose—sometimes quadrupling. The implications of this are significant, since the average citizen in an industrialized country spends 5 percent of his or her *life* (!) in an enclosed vehicle. The Canadian government tells me that I, as an average Canadian guy, can expect to live until about 80 years of age: That's 4 entire years of my life spent in a car. And for some of my fellow Torontonians who spend three hours a day commuting to and from work, I guarantee this number would be much higher. Granted, not every car will have the air quality issues of a new Chevy Tahoe. But many will.

Figure 12. VOC increases in Jeff and Rick after eight hours in a warm, new car (all units: mg/L)

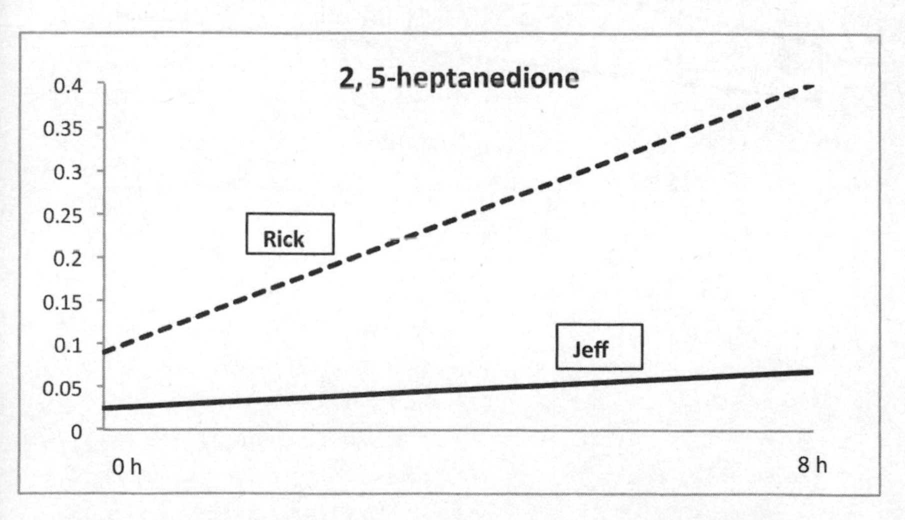

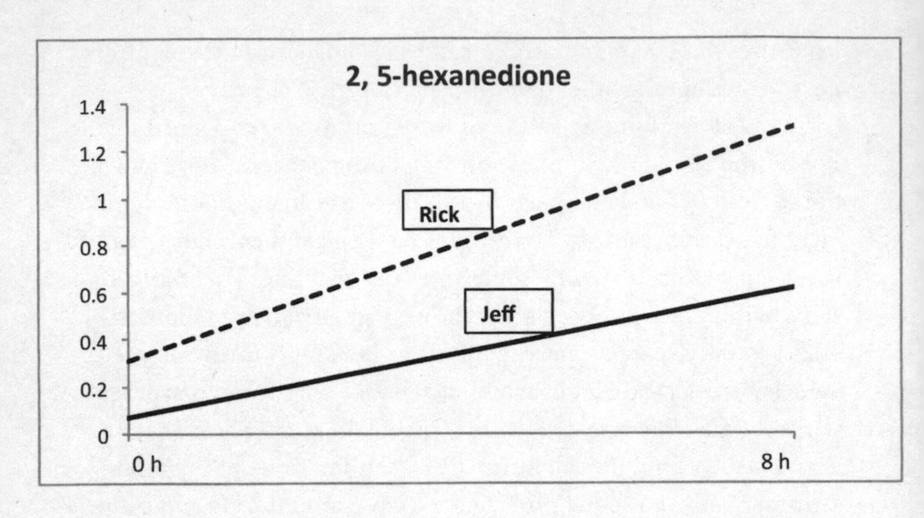
2, 5-hexanedione
1.4
1.2
1
0.8
0.6
0.4
0.2
0
Rick
Jeff
0 h
8 h

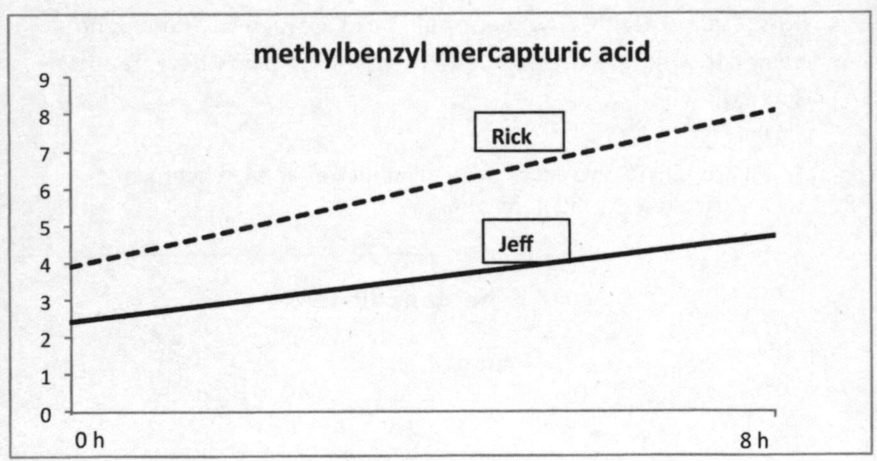
methylbenzyl mercapturic acid
9
8
7
6
5
4
3
2
1
0
Rick
Jeff
0 h
8 h

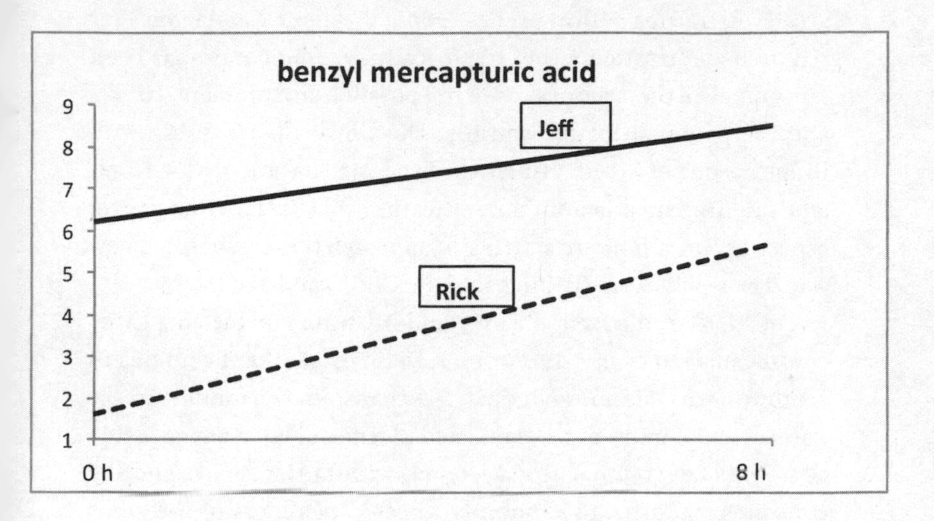

Our experimental result in a nutshell was that with every breath they take, millions of people around the globe, every day, are slowly filling their bodies with benzene and other unpronounceable toxic substances as they take a trip to the grocery store, pile the kids in the car to take them to soccer or head off for a summer vacation. We've included some specific conclusions regarding what to do about this in the last chapter of the book. Suffice it to say for the moment, if you have a new car, rolling the windows down and letting it air out is a good place to start!

"Toxic Soup"

One way to deduce how these sorts of pollutants impact us is to examine their effects on people who work with the stuff every day.

Though occupational exposure to chemicals and resulting health outcomes are poorly understood, a recent scientific study is noteworthy for its connection to automobiles. A team of researchers led by University of Windsor professor Dr. Jim Brophy, through the support of the Canadian Breast Cancer Foundation (CBCF), decided to investigate whether certain types of work posed occupational risks for increased rates of breast cancer.[10] The CBCF does a lot of great work raising both awareness about the disease and money for research and treatment, but historically the foundation has been fairly quiet on the issue of cancer and possible environmental links, and the potential for prevention. The CBCF's partnership with Brophy is one of the first times they've been so vocal on this front, and this initiative is another indication that large, mainstream cancer organizations are starting to approach these issues in a new way. The results were striking, and the CBCF validated them.

Let's look at the facts. Brophy and his team conducted a case-control study of over 2,000 women in Essex and Kent counties in southwestern Ontario—an industrial expanse known for its extreme flatness, automotive parts plants and plastic-manufacturing facilities. The case group of 1,005 women with breast cancer and the control group of 1,146 randomly selected members of the same communities were asked to provide detailed information about their occupational and reproductive histories. Women were classified by their levels of likely exposure to the chemicals in question and also by their occupations—farming, plastics manufacturing, restaurant or beauty work, for example—and the link between occupation and breast cancer rates was examined.

Here's the simple but powerful result: Industries characterized by high levels of occupational chemical exposure showed increased risk for breast cancer. More specifically, women in the plastics and food-canning industries were at approximately five times more risk of developing breast cancer before menopause. Women who worked in farming were about 36 percent more likely to develop breast cancer than the women in the control group. Women in tooling, foundries and metal-related manufacturing were at 73 percent greater risk. And increased risk for breast cancer was found in industries where women were exposed to the chemicals discussed in this book: BPA, phthalates, VOCs and others.

The results of Brophy's study are especially worrisome for women who work in automotive plastics plants. The plastics industry is characterized by a very high concentration of female workers; in Canada, for example, 37 percent of the plastics workforce is female—a higher proportion than any other industry in the manufacturing sector.[11] In Brophy's geographic study area, Windsor-Essex County, where the majority of plastics products are produced for the automobile industry—women make up over half the plastics workforce.[12] Similar proportions occur in plastics-related industries in the United States.

The plastic-manufacturing work done by the women in Brophy's study involved mostly injection moulding, which can be described as follows. After a product is designed—any product really, but in the automotive case, let's say a dashboard or the console between the seats with those handily designed cup holders—a mould is made to form the features of the plastic part. The injecting material for the part is fed into a heated cylinder, mixed around and forced into the mould cavity, where it cools and hardens.[13]

What does this mean in terms of the female workers' exposure to chemicals? The molten mixtures that become the final product are made up of resins, multiple additives and sometimes lamination films. The heating process emits vapours and mists into the plant that can include plasticizers, UV-protectors, pigments, dyes, flame retardants and unreacted resin components, among other things.[14] Many plastics release estrogenic chemicals, and additives like phthalates and polybrominated diphenyl ethers (PBDEs) in the molten mixture are known endocrine-disrupting chemicals. Furthermore, some of the ingredients present in the manufacturing of polymers (such as BPA and vinyl chloride) are carcinogenic.[15] So it's no surprise that Brophy and his team discovered that women in the plastics industry had double the risk of developing breast cancer as compared to the control group. What *is* surprising is that other studies have shown men in this industry to be at four times the risk for breast cancer as well.[16] And this adds weight to Brophy's findings for female automotive workers.

When I interviewed Brophy about the study, I was curious to know how these chemicals actually get into workers' bodies. Are they inhaled, absorbed through the skin or what? "These people are getting exposed in just about every way imaginable," he responded. "It's very important to emphasize that they're not just getting exposed to one toxic substance. They're getting exposed to the most unbelievable mixture of substances, all of which are known to potentially pose a threat to their health. And we know that when you get exposed to these mixtures, it's even more potent than if you get exposed to just one thing on its own." He told me that during interviews, his team repeatedly heard complaints of improper ventilation, and one of the women working with automotive plastics likened being in her workplace to being "in a toxic soup."

After sitting with Jeff in a car for eight hours, there was a noticeable effect on both our states of mind and our body chemistries, but studies like Brophy's show that the people who deal with high levels of these chemicals every day are at *significant* risk of truly terrible health outcomes. They are "canaries in the coal mine," warning of possible effects on the rest of us even at much lower exposures. Their stories and their illnesses should also make us reflect on the tangible impact our society's consumption patterns have on the workers that sustain them. The same occupational poisoning is experienced by workers worldwide: in vinyl-producing factories in China,[17] on farms in Sinaloa, Mexico, where workers (many of whom start as young as 10 years old) are exposed to pesticides,[18] and in shoe manufacturing in Brazil, where workers are exposed to n-hexane.[19] Is it fair to put their health at serious risk so we can keep using these chemicals in our products and lives? (Bruce investigates some of these issues—and the question of detoxing our economy—in Chapter 6.)

Strange Symptoms

In this chapter we've focused on VOCs, but other chemicals cause big problems when it comes to indoor air pollution—the ones that can enter our bodies in ways other than inhalation. Brominated flame retardants (flame-retardant chemicals that contain the element bromine) illustrate this very well. We looked at these chemicals in some detail in our first book, so let's do a quick refresher on why flame retardants, specifically the brominated kind, are such a problem. For starters, they're everywhere. Polybrominated diphenyl ethers, or PBDEs, are used in a wide range of common products, including building materials, electronics, furnishings, road vehicles and airplanes, plastics, polyurethane foams and textiles. They are present in a huge number of items that we all encounter day in and day out.

PBDEs are banned in Europe, and Canada has taken some action on them as well, particularly when it comes to their use in children's products (such as toys and pyjamas). In both animal and human studies, they have been shown to cause reproductive, thyroid, endocrine, developmental and neurological disorders, including decreased fertility, birth defects and hyperactivity.[20] To make matters worse, these flame retardants are persistent and bioaccumulative, having similar chemical structures to other pesky chemicals like PCBs (a formerly common chemical used in electrical equipment as a coolant). As a result they're found throughout the natural environment and in wildlife.[21] PBDEs are present in most human populations, and there are indications that despite some regulatory action, their levels in our bodies are continuing to rise.[22] Flame retardants have joined the ranks of chemicals on notice when it comes to long-term environmental and human health effects.

Our friend Miriam Diamond, a super-dynamic professor and researcher in the University of Toronto's Department of Chemistry, has done some impressive research on flame retardants in everyday environments. In 2004 she and her colleagues collected the grime from interior and exterior household windows throughout Toronto and surrounding areas to measure their relative levels of PBDEs. The results were compelling, showing significantly higher levels of PBDEs on the interior films—especially in more urban environments.[23] This finding prompted further investigation about interior PBDE levels and how humans are exposed to them, after which Miriam and her team published a groundbreaking study: the first exposure assessment of PBDEs indicating that house dust was the likely main route of exposure for individuals with high levels of the chemical in their bodies.[24] Miriam's research earned her the "Scientist of the Year" award from *Canadian Geographic* as well as the honorific "Dust Doctor" from her students. I phoned up the good doctor to ask her a few further questions about PBDEs, dust and indoor air quality.

"If PBDEs are ingredients in seemingly inert items like the foam in seat cushions," I asked, "how do they get into our bodies? Is it as straightforward as the fabrics and foams containing flame retardants breaking down over time and becoming dust that we inhale?" While this could be true of some older, ripped pieces of furniture (like the sofa in the student house I lived in during my undergraduate university days, which we got from the Sally Ann and which slowly yellowed and disintegrated before our eyes), Miriam explained that it's actually more complicated than that. She thinks there are two main ways PBDEs get into dust. She likened the first way to taking a hot shower and watching the water vaporize and condense on the cold window. "A product—like your computer or TV—heats up when you use it. As a result of heating, the flame retardants de-gas, or move from the plastic into the air. But PBDEs don't want to stay in the air because of the low vapour pressure and also because the air is colder than the electronic product. So then they partition into stuff, stuff like the greasy film on walls and floor and plant surfaces—whatever surface is available for condensation."

Another avenue for PBDEs entering household dust involves humans. Recent research by Miriam's colleagues shows higher concentrations of PBDEs on the bottom of the hand versus the top, indicating that the chemicals are getting onto our palms when we touch them. "We know that bits of chemicals come off the surface of objects that contain them," Miriam explained. "And we know that because all you have to do is touch something containing a flame retardant, like a computer, and you're going to get it on your hand. A little bit of the plastic or fabric seems to come off just by microabrasion." She went on: "Here's another scenario: You've got dust on your couch, and when you sit down, some dust floats up and is resuspended in the air. While the dust was settling on the couch, it had the opportunity to pick up the flame retardant from the couch, and when it's resuspended, it moves into the room. So clean dust goes down; contaminated dust goes up."

A study published in the spring of 2013 reminded us that it's not only dust in the home that we should be worried about. Researchers at Duke University looked at a new-to-the-market flame retardant known as V6 and its concentrations in foam baby products and household and automobile dust.[25] With the slow phase-out of PBDEs, V6 has been developed as an alternative, so it's no surprise that the newer flame retardant was found in high concentrations in the baby products and dust that were examined. What's more, the study found that concentrations of V6 were significantly higher in car dust than in house dust—which could mean that this flame retardant is used more in cars than in household products like foam mattresses and seating.[26]

When I think of dust I tend to think of dust bunnies. But it seems that toxic dust in your home or car is actually more likely to be distributed in places that you wouldn't normally expect. In short, it's not the dust bunnies that Miriam worries about. "Dust bunnies are what we call the reservoir dust. [But] it's *active* dust, the stuff in the centre of the room that gets resuspended in the air, that's the concern." Dust bunnies are usually pretty large, chunky particles as opposed to smaller bits. Think of them as dust carcasses: dead creatures mouldering away on the ground that won't bother anyone ever again. In contrast, the living dust, the really small and sneaky particles that we inhale and often eat (the motes visible in the sunlight as it streams through your kitchen window) actively move around.

In the same way that biologists track animals with radio receivers, Miriam Diamond tracks dust—and with equal zeal. Her work has revealed much about how these chemicals start out residing in the seat cushion under your butt and then go on to be an invasive pollutant in your bloodstream. And once PBDEs are in us, bad things start to happen. Miriam referred to a recent epidemiological study examining the effects of prenatal PBDE exposure and indicators of neurodevelopment in children 1 to 6 years of age.[27] The study recruited 210 women in Manhattan who were pregnant during the 9/11 attacks and subsequently delivered at one of three downtown hospitals. Each baby's umbilical blood was taken at birth and analyzed for the presence of toxic chemicals. At 12, 24, 36 and 72 months of age, various types of developmental tests were administered, and scores were compared to known developmental norms. The research team hypothesized that prenatal exposure to toxins, particularly PBDEs, emitted from the World Trade Center buildings after the attack would affect neurodevelopment in children. They were right: The cohort study reported lower scores on mental and physical development and lower IQ in infants and toddlers, especially in children whose umbilical blood had the highest concentrations of PBDEs. "It's weird," Miriam said. "As a mom, when I read the adverse effects, it just felt uncomfortably close to my circumstance. I thought of all those years working in front of a computer prior to and during pregnancy." She has always sent her children *outside* to play, and thinks that maybe we should *all* play outside more.

Allergic Generation

Indoor air quality researchers have also focused on another family of chemicals: our old friends the phthalates.

Not content to enter our bodies only through the personal-care products they infuse, phthalates also comprise a significant chunk of the dust in our homes. A particularly convincing series of studies has linked phthalate exposure to asthma and, perhaps surprisingly, the development of allergies.[28] One of the key researchers in this area, Carl-Gustaf Bornehag of Karlstad University in Sweden, told me that when he started his pioneering research, he didn't have phthalates in mind at all. "In the 1990s, in Sweden, we were collecting scientific data on environmental issues related to asthma and allergies, and we were focusing on moisture-related problems in buildings." Bornehag noticed that something was happening in damp buildings resulting in higher rates of asthma and allergies in children—and he wanted to know why.

So he started one of the world's biggest studies, called *Dampness in Buildings and Health*. It includes 40,000 kids and has tracked their lives and their health status in five-year increments since 2000. As Bornehag and his colleagues started to put together their mountain of data, they encountered a problem: No matter how hard they sifted, they couldn't find a relationship between moisture related exposures and asthma and allergies. For a while, they were stumped. Clearly, they were measuring the wrong things, so they kept analyzing and re-analyzing their air and dust samples, looking for some new trends. Finally, in 2004, they found the common thread: It was phthalates. The more phthalates in household dust, the higher the rates of asthma and eczema in kids. Given that a great deal of flooring in Sweden contains polyvinyl chloride (PVC), which is very rich in phthalates, the concentration of the chemical in household dust in many cases was through the roof. "It's been quite a journey," said Bornehag. "We looked for something, we couldn't find it, but we did find something else."

Bornehag is now expanding his study to another eight countries around the world and a whopping 100,000 children. It's slow going, but similar results linking phthalates to respiratory and skin problems and allergic reactions in children are popping up in other jurisdictions.

Scientists are careful not to overstate their conclusions, and Bornehag is no exception. He did, however, end our conversation with an intriguing inquiry: "There's no question that there has been a significant increase in allergies in people during the last five or six decades. And if you take not only phthalates, but a number of these endocrine-disrupting chemicals, they have all been introduced during exactly the same time period." It's not scientific proof, but certainly a compelling observation.

Observation, it turns out, might be all that is needed for solving some problems outside the realm of academe. An innovative community initiative from Children's Hospital in Boston occurred when medical staff noticed a common denominator among patients. Doctors there were seeing an unusually high number of asthma patients, the leading cause of admission at this hospital, from certain low-income neighbourhoods in the city.[29] They knew that asthma attacks can be set off by things like dust, mould, polluted air and other environmental aggravations common in low-income neighbourhoods, and they wondered if they could do something about it. The Community Asthma Initiative was born.

The programme identifies what they call "frequent fliers"—children who make frequent visits to the hospital's emergency ward, suffering from asthma attacks. Willing parents of these kids are then visited by a community health worker, who helps identify and educate the families about problem conditions and products in the indoor environment that can trigger attacks.[30] Visits from the community health worker mean going through a checklist of asthma triggers and indicators of problem environments—things like mould, cockroaches and mice. The health worker then asks the families some further questions in order to help calculate an "Asthma Control Score"—questions such as how often asthma has kept the child from school and how frequently the child or parent has noticed shortness of breath. Through the Community Asthma Initiative, children are increasingly scoring higher, meaning their asthma is better controlled, and the programme has been successful in helping families achieve "healthier homes" and healthier kids.

The success of the programme is being noticed in hospitals too. A report published on the programme noted "a 56 percent reduction in patients with any emergency room visits, and an 80 percent reduction in patients with any hospitalizations."[31] All this has been done through simple and low-cost changes such as dusting and vacuuming more often, avoiding chemical-heavy cleaners and minimizing the use of fragranced products like room fresheners, which can all aggravate asthma. It's encouraging to see improved quality of life by making the connection between toxic chemicals in the home and health conditions like asthma. The onus is still very much on the families to avoid the everyday products that contain these chemicals, but there is much to be gained in this. Who knew that a simple spring cleaning could have such tangible detox benefits?

Good News, Bad News

I had ploughed my way through a tonne of papers on hundreds of different indoor air pollutants, and my head was hurting. I needed someone to give me the bottom line. For an overview of what's been getting better and what's been getting worse when it comes to indoor air quality, I turned to one of Bornehag's occasional co-authors: Dr. Charles Weschler, a researcher at the University of Medicine and Dentistry of New Jersey. Recently, Weschler has penned a refreshingly clear "everything you wanted to know but were afraid to ask" treatise entitled "Changes in Indoor Pollutants since the 1950s."[32]

"In terms of many known or suspected carcinogens, at least in the U.S. and Canada, things have improved significantly," Weschler explained. For benzene, formaldehyde, asbestos and radon, the levels in typical homes and offices today are much lower than they were in the 1960s and 70s. "Something like radon wasn't even on the radar screen in the 60s and 70s," Weschler said. Some of the decreased levels have been very striking, and he cited two large surveys of VOCs in the indoor air of U.S. homes, one conducted between 1981 and 1984 and another between 1999 and 2001. "When you compare the two surveys, the average level of benzene went down from about 14 micrograms per cubic metre to about 2 micrograms per cubic metre. And we saw similar declines for other aromatic solvents: toluene, xylenes and ethylbenzene. So that's good news." In terms of carcinogens, the indoor environment is significantly better today than it was 40 or 50 years ago.

Weschler's assessment of "classic toxic gases" like carbon monoxide (CO), sulphur dioxide (SO_2), and nitrogen oxides (NOx) is similarly upbeat. "Fewer people are dying of carbon monoxide poisoning today, per capita, than were dying 40 or 50 years ago, due to a combination of factors," he said. "We have CO detectors in homes, but more importantly, CO emissions from a lot of combustion appliances are just much lower than they used to be." He pointed to the car as an interesting example of this change. "It's almost impossible today to commit suicide by putting a vacuum hose on your tailpipe because the CO levels have become that low. You can make yourself sick, but it's pretty hard to kill yourself." Similarly, levels of sulphur dioxide are down due to reduced use of sulphur-containing fuel and coal. And nitrogen oxides indoors are way down due to the most innocuous of changes. "Do gas ranges in Canada still have pilot lights?" he asked. Not many, I told him. "Right!" he continued. "So pilot lights were this continuous source of nitrogen oxide that could actually reach elevated levels in a tightly sealed home. No more."

Heavy metals are another good story. "Lead, mercury, cadmium—they're all much lower in homes today and you collect dust samples today and compare the levels to what you would have found 20 or 30 years ago and again you see orders of magnitude reduction. Lead and mercury used to be used in paints to prevent mould and mildew—same with cadmium. And as you know, that's no longer permitted for indoor paints. So that's good news." The last rosy update Weschler underlined was the change in pesticides and the fact that many of the most toxic ones have been curtailed. "You no longer see DDT or chlordane or chlorpyrifos or myrex used indoors."

However, the progress report regarding other chemicals is not so good. Weschler has collaborated with Bornehag on some of his experiments and shares the Swede's concern regarding the link between phthalate exposure and increased allergenicity. He detailed a large ongoing experiment they're conducting to get to the bottom of how and where people become exposed to different phthalates. The experiment involves analyzing for five phthalates (DEP, DiNP, DiBP, BBP and DEHP) in the dust from the bedrooms of five hundred different children on the island of Funen in Denmark. Weschler and his colleagues also collected dust from the 150 kindergartens and daycare centres that these children attended and again analyzed for the phthalates. In addition, they collected urine samples and analyzed them for the metabolites of the five phthalates in question. They also estimated indoor exposures for each of these phthalates, given what is known about their concentration in the dust and the air, and making assumptions about the rate at which the children were breathing and the amount of skin that was exposed.

The results were fascinating—and wildly variable, depending on the specific phthalate. "Only 5 to 10 percent of the DEHP found in the human body can be explained as a consequence of indoor exposure; the remaining 90 percent is presumably diet and mouthing on toys and similar pathways," he told me. "But for something like DiNP or DiBP, the indoor exposures that we've calculated can explain 40 or 50 percent of what we're finding in people's urine. And for DEP, indoor exposures are responsible for an even greater fraction of body levels of the chemical." Weschler's overall conclusion is that for many chemicals we find in our urine or blood, indoor exposures are making a significant contribution to our total exposure. "People do understand that diet, food and drink play a role in exposing them to chemicals, and many understand that personal-care products play a role—but the idea that the dust or air in the home could be playing a role doesn't register for many people."

Weschler is also interested in focusing specifically on the extent to which our bodies' chemical burden is a result of the indoor environment and exposure through the dermal pathway (that is, our skin). In his 2012 journal article "SVOC Exposure Indoors: Fresh Look at Dermal Pathways," Weschler examined the concentration of several semi-volatile organic compounds (SVOCs)—a category that includes common chemicals like BPA, phthalates, pesticides and flame retardants—in indoor air and on indoor surfaces. Using these levels, he then developed equations to estimate the levels of SVOCs that got through the skin of experiment participants.[34] What he found was that air-to-skin transdermal uptake can be comparable to, or larger than, inhalation intake for many SVOCs of current or potential interest indoors.[35] His results indicate that exposure to these indoor chemicals through the skin has previously been underestimated and that this pathway should be given more weight by exposure scientists and health officials when examining exposure to indoor SVOCs.

Why is this particular type of transmission SVOCs such a concern? First, the nature of the dermal pathway means that the dose of chemical entering the body is potentially much higher than for a comparable amount of chemical that is inhaled or ingested. Whereas inhaled or ingested SVOCs have to pass through various systems (respiratory and digestive) where their appearance can trigger defence mechanisms of various kinds, SVOCs passing through the skin can go *directly* to the blood and organ tissue. Secondly, in his treatise on the current state of affairs of indoor chemicals, Weschler makes the point that VOCs evaporate quickly—"Emissions from a product tend to decrease sharply during the first few weeks or months of a product's life"—but that this is not the case for SVOCs. SVOC emissions continue throughout the life of a product, and once absorbed by a surface (such as the skin), they tend to linger and can continue to absorb, even after the host surface (the product that was off-gassing the SVOC) has been removed.[36]

So it's no wonder that human exposure to these chemicals indoors is a serious concern. Weschler makes the case for improved biomonitoring for these indoor pollutants. That is, ongoing measurement of levels of these chemicals in the human body. This will certainly help illustrate the dimensions of the off-gassing problem, but as always, we also need to make our homes and other indoor environments healthier through our own efforts.

Industry Responds

I can still remember what my hard-bitten Newfoundland grandmother used to say about latex paint. "Pah!" (dismissive wave of the hand) "That stuff never lasts, b'y. Oil is what you need!" But what she might have thought in the 1980s isn't the case in 2013. In fact, as Carl Minchew, Director of Environment, Health and Safety at Benjamin Moore Paints, told me, you'll be hard pressed to find anything other than latex paints now. And this is a good thing.

Those older oil paints, Minchew said, were really solvent paints: 50 percent of the volume of the paint in the can was a solvent mixture that evaporated as the paint dried. Given that latex paints dry more quickly, don't make your eyes burn and can be cleaned off your hands with soap and water, rather than paint thinner, it's no surprise that over the past few decades, latex has enjoyed a rapidly increasing market share. The difference between VOC levels in my grandmother's standby oil paint and modern zero-VOC paints like Benjamin Moore's "Natura" brand "is almost infinite" according to Minchew. The early latex paints contained about 10 percent of the VOCs contained in oil paints, and with improving paint technology, this proportion is now down to virtually zero. "Natura," according to Minchew, is simply "a better-performing paint: It goes on better, dries better, hides better and is more durable under use." And it doesn't fill your house with noxious pollutants. What's not to like?

The market's transition from oil to latex paints is just one example of the ongoing revolution in the way that building materials are made. Whether your house paint is smelly or not is a fairly tangible and obvious thing, but many recent advances are far subtler, though no less important.

In North America the introduction of the LEED (Leadership in Energy and Environmental Design) certification system marked the turn toward better building materials. LEED is a third-party certification programme for the design, construction and operation of high-performance green buildings.[37] It provides building owners and operators with a specific framework that allows them to identify and meet practical and measurable green building design, construction, operations and maintenance solutions that will help contribute to a healthier overall environment.

I spoke to Thomas Mueller, President of the Canada Green Building Council, the group dedicated to promoting LEED in Canada, about the extent to which LEED is now paying attention to indoor air quality. He explained to me that within the LEED system, there are five main credit categories, and one of these is called Indoor Environmental Quality. The core of this category is related specifically to the products that are being used in a building. "These are usually what we call finishing materials," Mueller told me, meaning things like paints, varnishes and sealants, "and they tend to off-gas harmful pollutants during their life in the building."

Mueller said it was the 2000 version of LEED (the second version in the United States, the first in Canada) that really broke into the marketplace and became popular. "The 2000 version focused on and addressed in a more consistent way the idea that some of these products and chemicals needed to be avoided, to enhance the health of building occupants," he explained. Mueller notes the impact LEED has had by way of the paint example. "When paint first came up in LEED, there was a certain threshold for VOCs in paint. If you want these points [the Indoor Environmental Quality points] in LEED, you cannot use paints that have a threshold above so many milligrams of VOCs per litre of paint." These LEED standards introduced the idea of low-VOC paint to the wider marketplace for the first time, and since then low-VOC paints have increased in quality and availability in the building sector. Now carpeting can also be rated according to Indoor Environmental Quality points, and Mueller explains that this point system is being extended to other product categories as well.

Mueller is quick to point out that improvements in building materials can't all be attributed to LEED but notes that the power of the rating system allowed it to become a market mechanism. "The interesting thing is that these products—paint, carpet—are marketed commercially, and they're sold at a price point to be competitive in the marketplace with other products. All the paint manufacturers, all the carpet tile manufacturers, they all have to have these low-VOC products. And that gets rid of the premium. People just say, 'I'm not using this old stuff anymore,' and they don't buy it."

Since its inception in 1998 and the launch of version 2.0 in 2002, LEED's success is obvious: It's grown exponentially as evidenced by various metrics. According to the U.S. Green Building Council's February 2013 monthly report, 182,421 LEED projects currently exist around the world. In the United States LEED-certified and -registered projects represent over 10.5 billion square feet of space.[38] And since its launch in 2004, the LEED certification programme for existing buildings has also experienced explosive growth. In late 2011 the square footage of LEED-certified existing buildings surpassed LEED-certified new construction by 15 million square feet on a cumulative basis.[39] With LEED we have the chance to right some of our previous wrongs, and make sure that future buildings are safer as well.

At its inception, LEED was considered to be "aspirational" in terms of greening the building industry, but today it's considered to be a reasonable benchmark. A newer certification system, the Living Building Challenge (LBC), is looking to further transform the building industry with the idea of "nutrition labels" for building materials and increased transparency around what's in the products we use. I spoke to Jason McLennan, the author of the Living Building Challenge codification system and CEO of the International Living Future Institute (ILFI): a non-governmental organization that focuses on creating a world that is socially just, culturally rich and ecologically restorative.[40]

Jason is considered one of the most influential individuals in the green building movement today, and he's the recipient of the prestigious Buckminster Fuller Prize. But his more humble beginnings in Sudbury—an Ontario mining city known for both its toxic and its regreening experiences—are what really started him on his path. "I was inspired by my community and its environmental legacy," he told me, "and also its legacy of trying to heal the landscape and regenerate what we had so degraded. I was a participant as a young child in the regreening process and was connected to the environmental changes that occurred. When I went into architecture, I went into it explicitly wanting to make this my focus."

The philosophy of the Living Building Challenge arose out of Jason's graduate research and his work with "green" architects throughout the United States and Europe in the 1990s, but the actual rating system wasn't launched until 2006. Much like LEED, the LBC is a certification programme that addresses development at all scales. It consists of seven performance areas: Site, Water, Energy, Health, Materials, Equity and Beauty. These are subdivided into a total of 20 "Imperatives," each of which focuses on a specific sphere of influence.[41] An important aspect of the LBC is the idea of transparency and prevention of toxicity in the building industry. Through the LBC Jason and his colleagues have developed a Red List of materials and chemicals that builders may not use in any project that is seeking Living Building Challenge certification. The Red List also itemizes materials that should be phased out of production due to health concerns and known toxicity issues.

The Living Building Challenge is still in its infancy in terms of its scale and impact on manufacturers, but Jason believes the building industry is starting to take notice and respond. In May 2011 Google Inc. announced that it would be adhering to the Living Building Challenge's Red List and dropping any suppliers that were on the list. Considering that the company opens about 40,000 square feet of office space a week and that their latest project is slated to be over a million square feet, it's a powerful announcement that will be sure to make vendors listen. [42]

When I asked Jason to describe his preferred green building universe, he wasn't shy to explicitly suggest that the system he's developed is where we need to go. "All buildings should be Living Buildings: net-zero energy, net-zero waste, net-zero water, carbon neutral and free of toxins. We have to have a completely renewable energy–powered world without pumping more carbon." Jason's basic point is that we shouldn't have to think so hard about these things. It should be easy to do good, rather than being easy to do bad. "Everything I've said sounds crazy," he acknowledged. "But it's all doable now. The only reason it sounds crazy is because we live in a paradigm that thinks the way we design and build now is 'normal.' And it's not."

Money Pit Musings

Much of this chapter was written while I was sitting on my living-room couch, listening to the banging and drilling of renovations beneath my feet. Our house in east-end Toronto is one hundred years old this year, and our basement was showing its age: mould appearing on the 1970s faux-wood wall panelling, water bubbling up through the concrete pad and lighting so poor our kids found excuses to avoid venturing downstairs. The *Toronto Star* newspaper we found in the wall was dated November 29, 1941 (with the headline "Sink Nazis in Arctic"), and the vintage porn retrieved from a hidey-hole behind the old basement shower stall attested to the last time anyone had touched the rooms in question. Over a period of about four months, we lived with our basement's former contents distributed awkwardly throughout the rest of our small house while our stalwart contractor, Alasdair, conducted some major and long overdue structural improvements.

As I was writing about indoor air quality, we were making related choices about new windows, carpeting, paint and furniture. Most of our decisions involved trade-offs of one sort or another. In order to secure windows with fibreglass, rather than vinyl, framing, we delayed construction an extra month. Instead of going with bare floors, as recommended by the dust-averse Miriam Diamond, we decided to carpet the cold (now waterproof) concrete pad. We did, however, use Interface "FLOR" carpet tiles made from recycled materials and free of VOCs as well as natural wool carpeting on the stairs. And we were careful to use low-VOC paint and carpet underlay, and despite some recent traces of problematic things in IKEA sofas I chose to believe their renewed corporate commitment to ban any brominated flame retardants from their products (a commitment that hasn't been forthcoming from other leading manufacturers).[43] The Månstad sofa bed in a lovely Gobo blue-grey hue fits in just fine.

In the real world, such trade-offs are the rule. But as a result of the "new car smell" experiment that Jeff and I undertook, I will never again take for granted the quality of the air in the places where I spend 90 percent of my life. In the summer, when the slush and sleet disappear from Toronto's streets (my wife Jen jokes that Benjamin Moore should market a new paint colour called "Winter Toronto Grey"), we plan to have the basement windows and new energy-efficient back door open as much as possible.

Bruce always likes to say that detox isn't a one-shot deal but a philosophy and a lifestyle—a statement that is entirely apt when it comes to managing the quality of the air itself. By purchasing greener products on an ongoing basis and supporting initiatives like LEED, the Living Building Challenge and the Ecology Center's car report cards, we'll all breathe a little easier.

SIX: CLEAN, GREEN ECONOMIC MACHINE

~ Bruce re-imagines economy ~

Water, air and cleanliness are the chief articles in my pharmacopoeia.

NAPOLEON I[1]

It's Not Easy Being Green

One line of questioning Rick and I often hear had me stumped: "What should I do with my old Teflon pan, and what happens to the toxic coating when I throw it in the garbage?"

I had to admit I had no idea.

We are so often told by our readers that we "made" them throw away their Teflon pans, BPA plastic containers, triclosan-filled personal-care products and countless other toxic items that I began to feel a certain professional obligation to figure out what happens to all of this stuff when it's discarded.

We've talked in previous chapters about why it's important to avoid exposure to toxic chemicals in the first place and about what can be done to help our bodies expel these chemicals once they're in us. In this chapter I'll be focusing on the importance of discarding synthetic toxins responsibly—though, of course, the best approach would be to stop producing these chemicals in the first place.

Why subject you to this chapter on garbage and manufacturing policy? Because of the tight connection between waste, toxic chemicals and you (check out Figure 13 for a simple explanation of this cycle). The waste we create winds up in landfills, and chemicals from those discards (like BPA, triclosan and flame retardants) then leach into the surrounding environment.[2] Pharmaceuticals and personal-care products tossed down the drain or flushed away can end up in biosolids and effluents from wastewater treatment plants.[3] These chemicals wind up in the environment (like the Great Lakes, for instance, where concentrations of some flame retardants have doubled in the last five years[4]). And they also make their way into the source water that's sent to our cities and homes and then comes through our taps and back into our bodies.[5]

Figure 13. The toxic chemical cycle

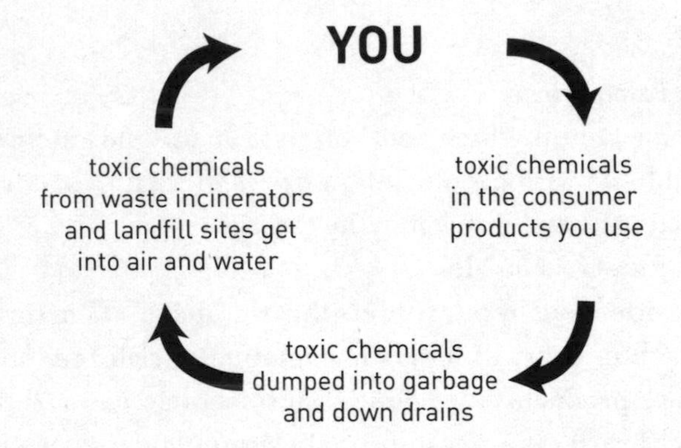

So when we detox our waste, we're detoxing ourselves.

The global economy (based, as it is, on the manufacture of goods that release harmful chemicals when discarded) also needs a detox diet—a really big one. Governments and businesses have been treating economic toxification with the equivalent of one-off diet fads and short-term cleanses, but these quick fixes are just as ineffective when applied to the economy as to individuals. And as the world falls off these one-time diets and starts to yo-yo, look out. You don't want to be standing near the global "gastrointestinal" explosion when it happens.

Greening the economy requires drastically rethinking what we produce and what we consume, right down to the chemical makeup of those items. It means taking a serious look at where our toxic trash ends up and how we manage to produce so much garbage in the first place, with so little thought about where it all goes. If we are to make any significant progress in reducing the volume of toxic chemicals being released into the environment every day, we'll have to answer the question of where those old Teflon pans go to die.

Trash of the Stars

I can imagine a new reality TV show. It would go like this. Contestants would each be given a giant garbage bag, and they would have to sort through it to figure out which movie star it came from, based on the trashy clues inside. Perhaps my location was going to my head: It was a spectacularly sunny day in southern California, and despite the thick yellow smog hanging over Los Angeles, I could just make out the Hollywood sign roughly 25 miles north.

I was standing on top of the Puente Hills Landfill, the largest landfill in the United States, a mountain of garbage over 200 metres high and covering nearly 2 square kilometres of land. It seems fitting that my environmental investigative work had taken me back to California, where conspicuous consumption collides head on with North America's leading environmental standards and the world headquarters for personal trainers, health nuts and detox fad diets.

My tour guide was Sam Pedroza, an Environmental Planner with the Joint Administration Office of the County Sanitation Districts of Los Angeles County (an independent agency, not part of the Los Angeles County government, Sam was quick to point out). Sam thoughtfully recounted for me the history of garbage in Los Angeles County. The sanitation departments were first established to manage wastewater in the LA basin in the 1920s and have since evolved into one of the largest waste management agencies in the world. Back in the 1920s Los Angeles was the 10th-largest city in the United States, with a population of just over half a million people. By 1930 it had more than doubled in size as oil and the American dream fuelled tremendous growth that continues to this day. Sanitation districts were established to better manage the privately run garbage dumps popping up all over the county. Most of the dumps were in LA's numerous valleys, which also happened to be sources of local drinking water.

Los Angeles's response to garbage is representative of the general history of trash throughout the industrialized world. A hundred years ago people didn't produce much garbage. Furniture, vehicles and building materials were made of wood and metal. People carried their produce and other items in sacks and pails and reused them. They grew their own vegetables and raised animals or bought their food from local farmers, and they didn't have electronic toys and devices that broke or became obsolete after a year. Anything they couldn't eat, compost or return to the soil was generally tossed into the wood stove or fire pit. And burning waste wasn't a bad thing because farmers and ranchers in the 1930s weren't burning plastic, mercury-filled batteries or Teflon-coated pans.

During the Industrial Revolution people started moving into cities and creating refuse at a rapid rate. With no farm fields to bury their waste in or open spaces where they could burn it, they tossed rubbish into rivers, down wells or into the streets, creating a serious public health issue. The nature of waste was also changing—from predominantly preindustrial organics and coal ash to early mass production of manufactured items such as iron and steel, glass, paper and fabric. Over the course of the 20th century, waste generation increased tenfold, from an equivalent of 42 kilograms per individual to 565 kilograms per person per year.[6]

It wasn't until the 1950s, when we first faced a "garbage problem," that we started to pay attention to waste disposal, Sam explained. As with much of LA's environmental progress, air pollution was the trigger. In the early 1950s backyard incinerators produced as much smog as cars, and water pollution from hundreds of little unregulated dumps in the valleys of SoCal was increasing. It was clear that the county needed to step in. But as Sam told me, in the 1970s the "garbage problem" hit LA especially hard, reflecting what was happening in most of the world. I attribute this problem to the "plasticization" of the economy. Within the two decades from about 1950 to 1970, paper bags, wooden toys, wooden furniture and cotton or wool fabrics were replaced by plastics and other synthetics. With the onset of "plasticization" came the disposable society—in large part because plastic items broke easily and could not often be repaired or were simply designed for single-use applications, such as plastic packaging. Sam identifies the mindset of "people just tossing out whatever they want" that led to overflowing garbage dumps and the need to create more and bigger landfill sites. More than 40 years later, with all of our composting programmes and recycling efforts, surely things must be getting better, right? Wrong. City dwellers today are producing almost double the amount of personal garbage as they were *just ten years ago*—from 0.64 kilograms of solid waste per person per day in 2002 to 1.2 kilograms per person per day in 2012.[7] The global trends of urbanization and plasticization are making it difficult for us to manage our colossal garbage systems failure.

In 2004, China surpassed the United States as the world's largest waste producer, and it is forecast to generate twice as much trash as the United States by 2030.[8] Most of us in Canada are convinced that with our world-leading recycling programmes, we are producing less garbage and more useful recycled goods. I'm afraid not. Plastic takeout plates, water bottles, food wrapping and plastic bags can all be recycled, but most aren't. What's worse, much of that plastic doesn't even make it into the garbage or if it does, it manages to escape, blowing off garbage barges or simply tossed into a ditch.

There's a floating island of trash in the Pacific Ocean called the Great Pacific Garbage Patch. It's hundreds of kilometres across and contains more garbage than any landfill site on earth. Ninety percent of this swirling mess is plastic. Marine debris is threatening not just oceans and lakes, but our supply of food and water as well.[9] In 2012 a group of scientists collecting water samples from the Great Lakes found up to 600,000 plastic pieces per square kilometre, almost twice as many as the highest count ever made in the Great Pacific Garbage Patch. Japanese researchers discovered that seaborne plastic pellets attract and concentrate harmful chemicals like PCBs; DDE (1,1-Dichloro-2,2-bis(p-chlorophenyl) ethylene), a breakdown product of DDT; and nonylphenols (breakdown products from detergents), which don't dissolve in water.[10] So in addition to the problem of plastics leaching, they also act as magnets for highly toxic pollutants, and these infused plastic particles are then ingested by marine animals. Scientists at the University of British Columbia's Department of Zoology examined 67 northern fulmars (foraging seabirds) on B.C.'s northwest coast and found that 93 percent of the birds had plastic in their stomachs, including twine, candy wrappers and polystyrene foam.[11] One of the birds had ingested 454 pieces of plastic.

Gulls *love* garbage. This was all too apparent when Sam drove me to the working face of the landfill, where massive bulldozers push the garbage and spread it out as it is dumped from truck after truck. Hundreds of gulls circled the freshly spread waste, largely oblivious to the "hawk cannon," a firecracker-like device that is shot out of a bazooka, whistling through the air and making the cry of a red-tailed hawk. The first one that went off worked better for me than for the gulls. In the 15 minutes I spent mesmerized by the operation, there must have been 40 huge trucks that pulled in, dumped their loads and left.

It all looked so easy. Dump the garbage; push it around; compact it by driving over it in machines with huge, studded steel wheels; scrape some topsoil onto it; and *voilà*: a budding recreation area! Sam compared the garbage mountain to an Aztec pyramid with graduated layers rising in almost perfect symmetry.

I remarked on the lush vegetation that had clearly been planted on the steep-angled slopes—no doubt well fed by the organic nutrients in the waste stream. Sam was pleased that I'd noticed that and was clearly proud of the facility. And so he should be, because it was the lushest vegetation I had seen in all of southern California.

I asked Sam to isolate the biggest waste problems Los Angeles faces today. "The changing face of garbage," he replied, explaining to me that plastic packaging is a much larger percentage of the waste now than the easy and more valuable recyclables such as glass, garden waste, paper, cardboard and aluminum, which are separated at the source for recycling. Just as I suspected. This was obvious, seeing the plastic bags and packaging being dumped, truckload after truckload. Sam mentioned the 3R waste management mantra, "reduce, reuse, recycle," but admits we are not really making much progress on the first "R."

e-Trash

Sam and I walked past a pile of old television sets and computer monitors in the Multi-Material Recycling Facility—which is largely a depository for electronic waste. "E-waste," Sam said, "is a major challenge." Apart from educational efforts to prevent people from discarding electronics, there didn't seem to be much else to do, he offered.

When it comes to the conspicuous consumption spelling bee, A is for Apple. Apple's ubiquitous i-devices are the epitome of trendiness and style. Every time the company introduces a new device, sales grow faster than those of the gadget it replaces. Within two years of its launch, the iPad exceeded 55 million units sold worldwide. Apple's CEO Tim Cook sums up the trend:

> This 55 [million] is something no one would have guessed. Including us. To put it in context, it took us 22 years to sell 55 million Macs. It took us about 5 years to sell 22 million iPods, and it took us about 3 years to sell

> that many iPhones. . . . And so, this thing is . . . on a trajectory that's off the charts.[12]

That's not the only trajectory that's off the charts. Developing countries are facing a tidal wave of e-waste as we seek cheap ways to dispose of defunct electronics that still contain valuable metals to be recycled. For instance, about 70 percent of the world's 500 million tonnes of yearly e-waste ends up in China.[13] The discarded TVs Sam showed me were next to bails of plastic bottles. Sam tells school kids that these recycled plastic bottles "will be your new toys from China." Plastic, paper, aluminum, steel and e-waste—even unsorted garbage from some cities—are destined for China where cheap and abundant labour makes hand sorting feasible.

A smartphone is replaced, on average, every 18 months,[14] and by 2015 over a billion smartphones will have been sold worldwide.[15] And they don't just sell themselves: In 2012 Samsung and Apple spent over three-quarters of a billion dollars on advertising campaigns trying to convince us to buy new ones. How much did they spend dealing with the e-waste from the phones they encouraged people to toss out?

Planned obsolescence, the practice of deliberately limiting a product's useful life, is one of the chief culprits of overconsumption. In the old days (and I can now recall those), a home telephone lasted about 20 years. My parents still own a functioning fridge that was built in 1952 (an ecological coup, inefficiency aside). But now we tend to expect products to either malfunction or become technologically obsolete within years, or even months, of purchasing them.

Printers are a case in point, and my printer is no exception. While attempting to print something last fall, a message appeared on the tiny screen saying, "Ink cartridge expired." I dutifully replaced the black ink cartridge (it cost $30) only to realize that the old cartridge was still nearly half full. Many printers are designed with a built-in "kill switch" effectively "timing out" once a predetermined number of pages is reached—not when the ink runs out.

Many major manufacturers require us to buy new ink cartridges when, in fact, there's enough ink left, in my experience, to print several hundred more pages. It appears that the business model for these companies is to discount the cost of the printer and to encourage buyers to purchase so many ink cartridges that after a couple of years, they will spend more on ink than on the printer itself. Thankfully the ink cartridges can be recycled at many office supply stores, but the point is that we shouldn't have to buy so many in the first place. Although I didn't see any discarded smartphones, iPads or plasma TVs at the landfill site, Sam assured me they were there and that we'd find some if we wanted to dig for them. We weren't too far from some of the wealthiest addresses in the world—along the coastline of Newport Beach and Laguna Beach. I imagined some local conversations along the following lines: "It's only an iPad 2 for heaven's sake. Throw it out" or "Are you kidding me? a 42-inch plasma TV? I wouldn't be caught dead with that thing in my house!"

China Syndrome

The region of Guiyu, China, has been in the media spotlight since the mid-2000s as the e-waste dump of the world. The area has been dubbed "the Chernobyl of electronic waste." In Guiyu an estimated 5,000 small businesses and 150,000 workers, many of them women and children, comb through our electronic refuse, separating the various components.[16] According to studies prepared by nearby Shantou University, women living there have an elevated incidence of miscarriages, children's bodies contain deadly amounts of lead and the soil contains the world's highest level of cancer-causing dioxins. Shantou professor Huo Xia has been measuring lead levels in the blood of local children since 2004. The 2010 test results showed the highest recorded levels to date, with 88 percent of children[17] exceeding the threshold for lead poisoning established by the Centers for Disease Control in the U.S. We all produce e-waste, and I'm no different. Somewhere in the back of my closets and drawer and in that old ice cream pail in the basement, I know I have old batteries, cameras, BlackBerries and lots and lots of adapters. When I take them to my local e-waste recycling drop-off, I've often wondered where they end up.

I contacted Cindy Coutts, President of Sims Recycling Solutions Canada, to find out. Sims Recycling is a top-notch e-waste recycling company, which has the capacity to extract and sort leaded glass, metal and plastic from their e-waste, maximizing recycling and minimizing the need to incinerate the leftovers or send them to landfill. With over 50 facilities worldwide, they are also one of the largest.

Cindy described the ever-growing and ever-changing mix of e-waste and what it contains: brominated flame retardants and heavy metals like mercury, leaded glass, cadmium, lithium, cobalt and radioactive materials—not to mention hundreds of different plastic compounds. Recycling them is a monumental technological challenge: The only way it can be done effectively is by hand. First, workers manually remove the most dangerous components, such as mercury switches and rechargeable batteries. Then a series of shredders, magnets, shakers, filters, blowers and belts separate the rest as best they can. Workers are typically standing by the belts, hand picking and sorting the items that the machines miss, making health and safety a pressing issue for recycling handlers.

One of the challenges Cindy has is making sure that the plastic they remove from e-waste is properly recycled. As with most e-waste, it is shipped to China, and one of Cindy's responsibilities is to find out where it ends up. Twice a year she travels to China for a firsthand inspection of plastics recycling facilities across the country. She described the scene in early 2013 of enormous black-and-white speckled mounds lining the roads and paths in village after village—like snowbanks going right up to people's doorways. "Plastic is everywhere. It's part of the landscape," she said. In one town Cindy asked about their waste management practices and was told that (incredibly) they had none. On her way out of town she could see evidence of this:

A bulldozer was pushing plastic garbage from the land straight into a river.

At another site where chemicals were used to help sort plastics, Cindy asked to see the wastewater treatment system they used. Officials pointed to a series of cement basins full of thick, dark water with bits of plastic floating around in them. It all evaporates, she was told, but finding this hard to believe, Cindy inspected the basins further and discovered a small hose at the bottom of each basin. She followed the hoses down to the river, where she saw the untreated chemicals emptying out.

Arriving at a recycling facility in yet another town, Cindy was welcomed by the workers, who were all dressed up to greet their special Canadian guest. She observed carefully their process of manually sorting plastics. One worker, in this case a young man, held a flame to a piece of plastic to see if it would burn. The plastics that didn't burn were tossed into a pile labelled "brominated flame retardants." For any that did burn, he waited a few seconds, blew the flame out and held the smoking plastic under his nose. He identified the type of plastic based on the scent of the toxic fumes and then sorted it accordingly.

Cindy wasn't in China to expose poor recycling practices or toxic waste dumping. She just wanted to find e-waste recycling companies that she could trust. Of the 70 preselected recycling facilities she visited, she approved 8 to be eligible to receive their plastic e-waste. How is it that manufacturers don't take responsibility for how or where their products are manufactured or disposed of, let alone for the toxic materials they use? Changes are happening in this area as manufacturers start subscribing to Extended Producer Responsibility (EPR) programmes. But how is this new waste mantra working out?

Exaggerated Producer Responsibility

Sam Pedroza told me that he sees Extended Producer Responsibility as the future of waste management. But others are not so sure. For instance, my close colleague and global EPR expert, Clarissa Morawski, says, "It's a great idea in theory. It basically makes companies responsible for their products at the end of life. Governments really like EPR too," Clarissa noted, "because along with shedding the responsibility of collecting and managing waste goes the costs of running EPR programmes. But this is where they tend to fall apart. The vast majority of EPR programmes are run by industry consortia that work collectively to implement the cheapest recycling option under the law."

Battery manufacturers are a case in point. Batteries can contain a slew of heavy metals such as cadmium, cobalt, lead, lithium and mercury, and their use is growing exponentially, even while the vast majority are thrown away and eventually end up in municipal landfills or incinerators. A voluntary battery collection programme that ran in Quebec for many years saw 85 percent recycling rates for the batteries, which were sent to a mechanical recycling facility in Ontario. Following the Quebec government's 2012 decision to regulate battery collection, the manufacturers organized themselves[18] and now send all the batteries they collect to a smelting facility in Pennsylvania, where only 25 percent of the materials are recovered; the rest are converted to a rock-like by-product called slag and used in road construction. It is a far cheaper option than the mechanical recycling—but with three-quarters of the material going to slag it hardly qualifies as successful recycling.

I spoke with Sam's colleague Mario Iacoboni, Supervising Engineer for the Los Angeles County Sanitation Districts, to find out more about the water and air quality at Puente Hills. I was curious to know whether anybody there ever measured (or even considers) what happens to BPA plastics, phthalates in discarded vinyl—or, of course, perfluorinated compounds (PFCs), the chemicals used to make Teflon pans and stain-resistant carpets. Mario stated that they follow government regulations and none of those substances are regulated, so they don't test for them. A simple lesson in why it's crucial to regulate toxic chemicals.

The Minnesota Pollution Control Agency found perfluorinated compounds in the leachate (the liquid that seeps out of landfills) and in the landfill gas (the gas emitted from the landfill into the atmosphere) at every Minnesota landfill site they tested.[19] PFCs are among the most persistent chemicals ever invented, meaning they do not break down easily in the environment and can build up in our bodies. The medical researchers examining the health effects of the PFC-contaminated drinking water in Parkersburg, West Virginia (described in *Slow Death*), have since found that PFCs in drinking water are linked to various cancers, high blood pressure in pregnant women, bowel disease, thyroid problems and childhood obesity.[20]

The Teflon pan question was still bothering me, so I asked some waste chemists what should be done with discarded Teflon wares. These chemists work at an experimental synthetic gas facility in Canada, where garbage is converted directly into inert slag and useable gas by zapping it at extreme temperatures. They use plasma technology, which (in simple terms) allows for a process that's like blasting garbage into oblivion with lightning. They gave my question some serious thought before getting back to me with an answer. Landfill, they said, was probably the best option, presenting the lowest likelihood of the non-stick PFC chemicals breaking down into more toxic by-products and getting into the environment. Incineration, as I suspected, was the worst option for PFCs, and one that should be absolutely avoided due to the temperature of incinerator fires and the highly toxic gases emitted. Plasma technology should be safe, they figured, because the plasma torches break down the molecular structure of the chemicals to simpler forms that are non-toxic.

Step one: Don't buy non-stick pans because of the associated health hazards.

Step two: Hope that your town doesn't incinerate garbage. If it does, drive to the closest town that landfills municipal waste, make sure it's garbage day and sneak your old pans into somebody's garbage can when they aren't looking.

I left the Joint Administration Office of the County Sanitation Districts of Los Angeles County with mixed feelings. I was genuinely impressed with the remarkable engineering feat I had witnessed at Puente Hills. But what about the non-stick pans, all of that e-waste, the plastic being sent to China and the Great Pacific Garbage Patch? Or the fact that even the most advanced Extended Producer Responsibility programmes are at best performing poorly and at worst creating the false impression of environmental progress while industry consortia opt for the cheapest recycling methods? Digging into waste, so to speak, gave me that "Houston, we have a problem" feeling. There was little evidence that we'd be able to keep our air and water safe by relying only on the waste end of the detox economy.

Detox Cleanse or Greenwash?

I strolled into the spiffy lobby of a Washington, D.C., hotel looking for signs that would point me to the 16th Annual Green Chemistry and Engineering Conference. I was directed up an escalator to the second floor, where I spotted a banner with a large red diamond and the word Dow in it. "This must be the place." I came to Washington in an attempt to understand whether we were on the verge of a green chemistry revolution or whether this detox cleanse for the economy was mostly a greenwash.

There were sessions on "recombinant cellulolytic bacillus subtilis," "methacrylated lignin model compounds as monomers" and "synthesis of bioactive 4-methyl-6-[(1-alkyl-1H-1,2,3,-triazol-4-yl)methoxy]-2H-chromen-2-one." Not knowing anybody or even where to start, I walked over to the information desk. "Is there a session called 'Green Chemistry for Dummies?'" I inquired hopefully. A pleasant older woman smiled and handed me a conference agenda and a pocket card with the Twelve

Principles of Green Chemistry printed on it (more on these below). Reviewing the meeting programme in greater detail, I was pleased to see a few familiar names: people like my close colleague Pete Myers, founder of Environmental Health Sciences, and Terry Collins, one of the world's green chemistry gurus. They were both presenters at the conference. I approached Terry after he concluded his formal remarks. Many people were vying for his attention, so I figured I'd better have one very good question for him. The perfect question hit me. "Hi, Terry, do you have dinner plans tonight?" Happily, he didn't.

Terry Collins is a chemistry professor at Carnegie Mellon University, where he has taught for 25 years and where he heads the Institute for Green Science. After seeing the destruction and health problems caused by pesticides and pulp mills in his native New Zealand, Terry set out to find a solution. He discovered an oxygen-based catalyst to replace chlorine in pulp and paper manufacturing. His quiet voice, soft New Zealand twang and friendly smile belie his stature as one of the world's most sought-after chemists and as the first professor of green chemistry. We had a fabulous meal of east coast fish (we chose the least toxic options), and I had Terry to myself for two hours of mind-bending conversation on everything from climate conferences ("hopelessly pathetic") to the petroleum economy ("it will kill us") to the booming gas extraction technology called fracking ("people are being seduced") and the role of universities ("a huge part of the problem"). And, of course, we talked about green chemistry ("a mutually reinforcing concept"). I asked three different people, including Terry, whether green chemistry was the "real deal" or whether it was susceptible to corporate greenwashing. They all had the same response: "It depends." Terry believes the philosophical underpinnings of green chemistry are very real. But he cautions that there's a huge risk of it being co-opted by the conventional chemical industry. "Take Dow [the sponsor of the conference], for example. They are expanding chlorine use and therefore producing more PVC [polyvinylchloride] and building up dioxins," he said. "Or consider ethane from methane in fracking gas, used to make PVC. People miss the ethane to ethylene to PVC plastics connection." He is referring to the American natural gas boom now underway where an environmentally contentious technique called hydraulic fracturing (fracking) is used to extract methane (natural gas), from which ethane is derived, followed by ethylene—a primary feedstock in the manufacture of PVC.

The Twelve Principles of Green Chemistry

The next day I attended several sessions and started to get the gist of green chemistry. In every session the Twelve Principles of Green Chemistry moulded the discussion as clearly as the Ten Commandments have shaped Western law. Here are the Twelve Principles:

1. *Prevention*. It is better to prevent waste than to treat or clean up waste after it has been created.

2. *Atom economy*. Synthetic methods should be designed to maximize the incorporation of all materials used in the process into the final product.

3. *Less hazardous chemical syntheses.* Wherever practicable, synthetic methods should be designed to use and generate substances that possess little or no toxicity to human health and the environment.

4. *Designing safer chemicals*. Chemical products should be designed to effect their desired function while minimizing their toxicity.

5. *Safer solvents and auxiliaries.* The use of auxiliary substances (e.g., solvents, separation agents, etc.) should

be made unnecessary wherever possible and innocuous when used.

6. *Design for energy efficiency*. Energy requirements of chemical processes should be recognized for their environmental and economic impacts and should be minimized. If possible, synthetic methods should be conducted at ambient temperature and pressure.

7. *Use of renewable feedstocks*. A raw material or feedstock should be renewable, rather than depleting, whenever technically and economically practicable.

8. *Reduce derivatives*. Unnecessary derivatization (use of blocking groups, protection/deprotection, temporary modification of physical/chemical processes [these are techniques used where certain kinds of chemicals are added to aid in the production of other chemicals]) should be minimized or avoided if possible, because such steps require additional reagents and can generate waste.

9. *Catalysis*. Catalytic reagents (as selective as possible) are superior to stoichiometric reagents [catalysts enhance chemical reactions and can often be recovered in the chemical process whereas stoichiometric reagents are consumed by the chemical reaction and cannot be recovered, making it a less efficient and more wasteful process].

10. *Design for degradation*. Chemical products should be designed so that at the end of their function, they break down into innocuous degradation products and do not persist in the environment.

11. *Real-time analysis for pollution prevention*. Analytical methodologies need to be further developed to allow for real-time, in-process monitoring and control prior to the formation of hazardous substances.

12. *Inherently safe chemistry for accident prevention.* Substances and the form of a substance used in a chemical process should be chosen to minimize the potential for chemical accidents, including releases, explosions, and fires.[21]

As one speaker put it: "Green chemistry needs to become an attitude, not a slogan." There were big ideas that sounded truly transformational and pragmatic proposals based on the Twelve Principles such as tidying up chemical manufacturing by producing less waste, using fewer catalysts and reducing energy inputs. Several speakers described new software programmes that predict the potential toxicity of newly invented chemicals. Certain physical properties of molecules may indicate the likelihood that they will be persistent, carcinogenic or mutagenic. Chemical characteristics can then be run through computer models before the chemical is synthesized, thereby preventing the manufacture of chemicals with a high likelihood of having undesirable properties. Sounds encouraging.

Cleaning Up the Chemistry Act

John Warner's name was all over the Green Chemistry conference. He is, after all, the co-author of the green chemistry "bible," where the Twelve Principles of Green Chemistry first appeared.[22] Warner has a PhD in chemistry from Princeton, 10 years' experience working as a chemist for Polaroid, 10 years in academia and 250 patents and research papers, and he now heads his own private research institute with 30 PhD chemists working for him.

I was struck by how John's credentials dwarf those of the

chemists who make careers out of dismissing the pollution concerns of environmental and health advocates and who defend the use of toxic chemicals. So I asked him why some chemists ignore the health hazards posed by low levels of synthetic chemicals and seem to dismiss green chemistry. "For the past 40 years universities have not taught chemistry students anything about toxicity—not until Terry Collins started at Carnegie Mellon," he said.

During his own chemistry studies at top universities, John never once had a class in toxicology and was never required to learn what makes a molecule toxic. That is why, he claims, so many conventional chemists don't understand toxicity—because they're working with scientific concepts from the 1960s and are oblivious to the fact that the world of chemistry has moved forward. I was reminded of the quotation from Nobel Prize–winning physicist Max Planck: "A new scientific truth does not triumph by convincing its opponents and making them see the light, but rather because its opponents eventually die." John told me that chemists are completely unregulated, unlike dentists or doctors or engineers, and they are not required to have a licence to practise. "Anyone can make a molecule that has never existed in the world before. Even if it's the most potent carcinogen in history, nowhere in their education are they ever required to have any knowledge of how to look at a molecule and say: 'Gee whiz, this might be a carcinogen. I might not want to make it.'"Changing the curriculum for chemistry students was an obvious and exciting challenge for John, so he created the "Green Chemistry Commitment" to accelerate learning. To date, 120 universities have signed on. His goal is to have all 600 universities in the United States that have chemistry departments commit to teaching toxicology and green chemistry within the next five years.

John narrowed down the Twelve Principles of Green Chemistry to five underlying principles that need to be followed for green chemistry to succeed:

1. Standardize toxic testing protocols.
2. Test products, not molecules.
3. Label products, not ingredients.
4. Establish timelines for companies to phase out toxic products.
5. Create a network of toxicology testing centres and train a workforce of testing technicians.

The first priority, John said, is to focus on the tests used to determine toxicity, not on individual chemicals. Manufacturers of chemicals and consumer products pass all the safety tests before a product goes to market, but different organizations use different safety tests with different assumptions, and they reach different conclusions. This needs to be cleaned up. John described priority number two: We need to move from testing molecules to testing products, and that might mean grinding up a smartphone, for example, and having it tested. In that way we can label the smartphone—not just a molecule—as containing a carcinogen or an endocrine disruptor. The third priority is labelling. Companies hide behind trade secrets as a way to avoid labelling ingredients, but if they are required to label products, that concern disappears (recall the hidden chemicals in hair products Rick described earlier). By indicating that *something* in the product is a carcinogen, people can choose to buy it or not. It takes the emphasis off the individual chemicals and turns the label into helpful consumer choice information about the product as a whole. Priority number four gives companies a set window of time to find nontoxic alternatives for toxic ingredients they are currently using in their products. The fifth priority is to create a toxicology testing infrastructure. Toxicology data exists on only about 400 chemicals of the 80,000 or so that are used commercially around the globe. Coupled with training and workforce development, this could become a politically attractive programme that focuses on skills development in the biotech sector, including improving the skills of community college students and displaced workers. Not only would workers be trained for entry (or re-entry) into the workforce, but they would have an immediate and useful function to play in supporting a stable, greener economy.

John's standards, as reflected in these priorities, are obviously high. But he's also pragmatic. "We need to be working *with* Dow, 3M and DuPont," he told me, "not *against* them. We've got to allow them molecular redemption and let them find a path forward that doesn't vilify them," he said.

John was frustrated that too many people want to focus only on the problems. "I can find infinite amounts of money to scare pregnant women, inject bunnies and publicize the dangers of some toxic chemical. But try finding money to invent a safe alternative and the same people say, 'That's not what we do.'" He lumps certain environmental groups and foundations together as having "romantic" ideas, with little hope of succeeding. "We need workable solutions to people dying and getting sick from toxic chemicals, not people trying to take down capitalism." Given the choice between being effective today in solving these problems or feeling really good about romantic notions of change and being completely ineffective, John would rather be effective.

Copying Nature

John and Terry are two parts of a green chemistry triad. The third is Janine Benyus, founder of the Biomimicry Institute and author of *Biomimicry: Innovation Inspired by Nature*.[23] Biomimicry, according to Janine, poses this question: What if, instead of manufacturing synthetic chemicals and harming Nature, we turned to Nature for help?

Biomimicry, much as it sounds, is the emerging field of modelling chemical and industrial design after the designs found in Nature. Janine described it more poetically: "We are asking Nature for inspiration." We are the students; Nature is the teacher. And here are some examples of biomimicry in action: the aerodynamic nose of a high-speed train mimicking the bill of a kingfisher, the super-slippery fabric of an Olympic swimsuit modelled after shark skin, and water collection nets (capable of extracting water from fog) designed to imitate the shell of a desert beetle. Janine noted that the root of so many sustainability challenges is the fact that we are battling the fundamental human urge to control and suppress Nature. Biomimicry, by contrast, gives Nature the lead. She scoffed at the idea of humans as stewards of the planet. "As if we know what we are doing. How well are we stewarding the world's fish population? Any chance we will be able to avert massive climate disruption? We should not even pretend that we have the understanding or frankly the motivation to be stewards; we are, after all, self-interested."

When I asked her to explain green chemistry from the perspective of biomimicry, she used Nature's medium for chemical experimentation to explain. "Life chose water, not solvents," she said. In the natural world chemistry is based on water, but industrial chemistry relies on high temperature, high pressure and petroleum-based toxic ingredients. "Nature, for example, does not add sulphuric acid to things," said Janine. "Heat, beat and treat" is Janine's refrain, referring to traditional chemical manufacturing techniques. The Holy Grail for green chemists is shifting from petroleum-based to water-based chemicals in manufacturing. Nature uses enzymes to facilitate "shape-based" chemistry, where the shapes in enzymes literally hold molecules together. I wondered if the action of the claw-shaped chelation chemicals that travelled through my blood (see Chapter 3) were an example of this process. "In green chemistry you can play with shapes—and physical things attaching together like puzzle pieces," she explained. It seems incredible, but putting the very same atoms in slightly different configurations—either bonded differently or aligned differently or with slight variations in relative proportions of carbon or hydrogen—can create fundamentally different substances. How, for example, can graphite—soft, dark and grey—be made of the identical chemical element that constitutes a hard, clear, pure diamond? It is largely the way in which the carbon atoms form chemical bonds.

Janine explained that in Nature, atoms line up and self-assemble into molecules. The process is somewhat like shaking a box of Lego pieces, pouring them out onto the floor and having the contents self-assemble into a castle. Nature makes sure that the "pieces" (the atoms) find each other in some medium (while they're swirling around in a pool of tepid water, for instance). Nature's chemistry relies on negatively and positively charged atoms being attracted to each other to produce stable chemicals and also on shapes being drawn to each other, Janine explains. Modern synthetic chemistry is much more about forcing atoms. There's a concept called "atom efficiency" in Nature, which means that atoms are not wasted, and it's this phenomenon that's described as "atom economy" in the second of the Twelve Principles of Green Chemistry. Most industrial chemistry requires the addition of many more chemicals than will end up in the final product. And this leads to vast amounts of wasted chemicals that may be discarded or discharged in wastewater. In Nature, however, there is no waste. The ingredients that go into creating a naturally occurring chemical stay in that chemical. Nothing comes out, so there are no by-products and therefore no toxic by-products. In addition, Nature does not add solvents, catalysts or other chemical agents. Biomimicry follows Nature's lead and therefore helps avoid the use of toxins in manufacturing.

Here's an example. If you get a receipt from a cash register that is shiny and coated with plastic, there's a good chance it contains the hormone disruptor bisphenol A (BPA). Janine and her green chemistry colleagues John Warner and Terry Collins were charged with figuring out a way to eliminate BPA from these little pieces of paper that clutter my desk. The lettering on a BPA receipt appears when certain parts of the receipt are heated, and this function is called "darkening during development." In biomimicry, the goal is to find out how Nature performs functions and then to imitate them. Janine turned to beetles for inspiration because the carapace (hard shell) of a beetle starts out as a pale liquid but hardens and darkens as the pupa changes into a beetle. If this were an industrial process, manufacturers would likely use some toxic solvents to create the substance. Then they'd expose it to high heat to cure it and harden it. Then they'd paint it to make it dark and probably add some kind of shellac or waterproofing to make it durable. Nature does all of this simply by having the chemicals self-arrange in water at air temperature. Over time the shell is created, and then it hardens and darkens and becomes incredibly strong, durable and flexible. But can these shell-darkening properties be mimicked so that non-toxic cash receipts can be made? The research is underway—led by John Warner's team. A new, non-toxic, water-based manufacturing process in the making.

Life Chose Water

What could be a simpler way of framing our future priorities as a species than basing them on the principle that Janine described: Life chose water. Because life chose water, we need to choose and protect water in order to safeguard our lives and the lives of our children and their children. Consuming adequate quantities of water is one of the most important things we can do to detox our bodies—but that's true only if we have access to clean, fresh water. Air, of course, is just as important to life as water, and if we can't maintain clean air and pure water, we have no hope of staying healthy—regardless of the detox protocols we follow. People often ask me whether I use a water filter at home. I do—even though I know that Toronto's drinking water is among the best municipal water in the world. Toronto's Water Quality Unit, for example, performs over 70,000 bacteriological tests each year on my drinking water, and it monitors over 300 potential chemical contaminants[24]—hazardous metals such as aluminum and zinc, dozens of poisonous organic compounds from ammonia to xylenes, among many others. But there's one exception: endocrine-disrupting (that is, hormone-disrupting) synthetic chemicals (EDCs), which are rarely included in any drinking water protocols. This is mostly because of the high costs of testing and the lack of analytical technologies and infrastructure needed to detect a wide range of endocrine-disrupting chemicals and their many metabolites. Therefore, most of the data on EDCs in drinking water come from targeted research projects, so we don't know the true levels of these chemicals in our water.

I take the quality of my tap water for granted. But so did the people of Milwaukee, Wisconsin, in 1993 and the residents of Walkerton, Ontario, in 2000. Milwaukee is a mid-size American city on the coast of Lake Michigan about 160 kilometres north of Chicago. Walkerton is a small Canadian farming town near Lake Huron about 160 kilometres northwest of Toronto. The two are about 500 kilometres apart as the crow flies, over two Great Lakes. Milwaukee is home to the largest single water-contamination event in recorded American history, when more than 400,000 people fell ill and over 100 people died from *Cryptosporidium* contamination.[25] *Cryptosporidium* is a nasty parasite that lives in the gastrointestinal tracts of many animals, including cows and humans.[26] It still isn't clear how it ended up in Milwaukee's drinking water—possibly through human sewage or possibly because of cows getting into a local river. Runoff from cattle manure was *definitely* part of the problem in Walkerton, which caused one of the most deadly drinking water incidents in Canadian history. In May 2000 the town's water supply was contaminated by a highly infectious strain of *E. coli* from farm runoff. About 2,500 residents became ill and 7 people died as a direct result, and many more have permanent kidney damage. An inquiry into the tragedy, known as the Walkerton Commission, laid most of the blame on improper operating practices and negligent behaviour of the Walkerton Public Utilities Commission. However, the Ontario provincial government was also faulted for cutting back on environmental monitoring, poor water-quality regulation and failing to enforce the existing water-quality guidelines.[27] The inquiry's report resulted in sweeping reforms of Ontario's water-testing standards.

These tragic events have also contributed to surging interest in home water filters. Sales of water-filtration systems in Canada shot up after 2000, as did sales of bottled water. In 1999 bottled water consumption in Canada was about 24 litres per person; by 2005 that number had jumped to 60 litres per person, with sales exceeding C$650 million.[28] The Milwaukee and Walkerton events awakened North American families to the quality of their drinking water, and for many others, they raised concerns about the adequate protection of water systems. Consider the very basics of the hydrological cycle. Oceans, lakes and rivers store water. The water evaporates and condenses, forming clouds, and falls to the earth as precipitation. Precipitation can appear in the form of rain or snow, and it may end up stored in a mountaintop glacier for centuries, or it may disappear into the soil of a farm in the Midwest in an instant. The process of evaporation and condensation, together with movement through rivers, soil and rock formations, cleanses and purifies water. Unless the process is interrupted.

Human activities are serious interruptors of this natural water cycle. We spray pesticides and herbicides, and we run factories, manufacturing facilities, coal-fired electricity stations, garbage incinerators and wastewater treatment plants. In North America the governance of water quality and safety is a shared duty, with states and provinces responsible for drinking water guidelines and municipalities that oversee water treatment facilities doing the day-to-day management and monitoring.[29] Governments issue water-quality limits, standards and/or permits, but many of these limits were established decades ago when populations were smaller, competition for water was not so fierce and the chemicals we used were not nearly as complex as they are today.

Despite all the reassuring work on the part of my city, by late summer, my tap water, which comes from Lake Ontario, smells a little swampy. This is the main reason why I also have my water filter. It does a great job of getting rid of odour, and it helps remove other contaminants that may be present. Water-quality guidelines, however, are designed to make sure that people do not actually become *sick*, and the threshold of safety for heavy metals, pesticides, pharmaceuticals and sundry other things is debatable.

Let's look at trace pharmaceuticals as one example of the contaminants people ingest from their tap water. A 2012 World Health Organization report noted that the last decade has seen a significant increase in the number of studies investigating pharmaceuticals in drinking water and concerns have been increasing about the potential risks to human health related to low-level exposure.[30] Early studies in the United States in the 1970s first reported the presence of active ingredients from heart medications, pain relievers and birth control pills in wastewater. Since then, peer-reviewed research has found between 15 and 25 different pharmaceuticals in treated drinking water worldwide.[31] And synthetic endocrine-disrupting chemicals like BPA and phthalates from landfill leachate and municipal sewage effluent may end up in tap water as well.[32] These chemicals are additives in plastics and are released from products as they degrade over their life in landfills. BPA and phthalates get into the leachate, and especially in landfills with no leachate treatment facilities, they're discharged into the surrounding rivers and groundwater, which can be sources of drinking water.

Some people don't use water filtration systems, but they still distrust tap water, so they look to other sources of pure, clean water—including bottled water. But few people stop to ask whether bottled water is, in fact, the safer choice. Bottled water has its own problems.

Bad Bottle

With images of snow-peaked mountains on the labels, we imagine bottled water gushing from a brook in the Swiss Alps into a little plastic container that somehow makes its way onto our grocery store shelf. However, much bottled water is "purified water"—nothing more than packaged municipal drinking water—the same water that would come out of your tap that's been put through a filtering system and sold to you at a high price. In Canada bottled water is actually regulated as a food product.[33] It's subject to monitoring for micro-biological content, but limited details of these analyses are provided.[34] Manufacturers are not required to test for trace toxic contaminants like those tested for by municipalities in tap water, nor are they required to report any testing analyses to any authority.

In 2011 Americans bought over 9 billion gallons of bottled water valued at US$22 billion.[35] The manufacture of the plastic bottles alone used up 17 million barrels of crude oil.[36] In North America only 13 percent of the more than 2 million tonnes of plastic water bottles are recycled—87 percent end up in landfills each year.[37] These bottles take centuries to decompose, and if they are incinerated, they emit toxic by-products that are released into the atmosphere. Finally, for a city with 1 million people, tap water costs approximately 80 cents per 1,000 litres, while bottled water costs, on average, US$527 per 1,000 litres, not including the enormous environmental costs.[38]

If there is one simple thing that every human can do to improve environmental conditions, it is to *stop buying bottled water*.

Filtering Out the Negatives

Removing unwanted contaminants from your drinking water is one of the simplest ways to detox your personal environment. Household water filters generally fall into two categories: point-of-entry units and point-of-use units. Point-of-entry units treat water before it is distributed through the house, while point-of-use units are self-contained, generally portable, devices that include countertop filters, faucet filters and under-sink units.[39] I have a point-of-use filter built into my fridge with a replaceable cartridge. At my cabin I have a countertop filter that filters lake water. I did a fair bit of research on my countertop filter, but can I be guaranteed that it is doing its job?

Many filters use more than one kind of filtration technology within their systems. NSF International is a widely recognized independent, not-for-profit organization that provides standards development and certification for the world's food, water, health and consumer products,[40] and as a general rule, filters labelled as meeting the NSF Standard 53 are certified to remove contaminants of concern in your tap water. Standard 53 addresses point-of-use and point-of-entry systems designed to reduce specific health-related contaminants, such as *Cryptosporidium*, *Giardia*, lead, VOCs, and MTBE (methyl tertiary-butyl ether), which may be present in public or private drinking water.[41] Generally, filters that meet the NSF Standard 53 are geared toward treating water for health and not just aesthetic qualities. The NSF certification programme is not flawless—and certainly doesn't cover all products—but it does provide assurance that at least some manufacturers' claims have been independently tested and verified.

You can see the results for specific types of filters are presented in Table 7, and the NSF International website provides a useful product database that can help you choose a filter that meets your needs.[42]

Table 7. Types of filters and their uses and mechanisms of elimination

Type of Filter	How It Works	Use	Elimination
Activated carbon filter	Positively charged and absorbent carbon in the filter attracts and traps impurities.	Countertop, faucet filters and under-the-sink units	•tastes and odours •heavy metals such as copper, lead and mercury •disinfection by-products •parasites such as *Giardia* and *Cryptosporidium* •pesticides •radon •VOCs like dichlorobenzene and trichloroethylene (TCE)

Type of Filter	How It Works	Use	Elimination
Cation exchange softener	"Softens" hard water by trading minerals with a strong positive charge for one with a lesser charge.	Household point-of-entry units	•calcium and magnesium •barium and other, similar ions hazardous to health
Distiller	Boils water and recondenses the purified steam.	Countertop or whole-house, point-of-entry units	•heavy metals like cadmium, chromium, copper, lead and mercury •arsenic, barium, fluoride, selenium and sodium
Reverse osmosis	A semipermeable membrane separates impurities from water (process wastes significant amount of water).	Under-the-sink units	•parasites like *Cryptosporidium* and *Giardia* •heavy metals like cadmium, copper, lead and mercury •arsenic, barium, perchlorate, nitrate/nitrite and selenium
UV disinfection	Ultraviolet light kills bacteria and other microorganisms.	Under-the-sink units (with a carbon filter)	•bacteria and parasites •class A systems protect against harmful bacteria and viruses •class B systems designed to make non-disease-causing bacteria inactive

Adapted from Natural Resources Defense Council, "Water Issues: Consumer Guide to Water Filters," accessed February 2013, http://www.nrdc.org/water/drinking/gfilters.asp.

The Wrong Rules

When I was running the Great Lakes Conservation Program at the Toronto-based Laidlaw Foundation, we looked at the connection between health and toxic chemicals in drinking water from the

Great Lakes. The Great Lakes Basin is the source of drinking water for 40 million North Americans,[43] and research was mounting about the harm that chemical discharges into the watershed were doing to birds, fish and the people who relied on the Great Lakes for their tap water.

That was 20 years ago, when it was uncommon for people to make the connection between human health and the condition of the environment. That situation has changed, fortunately, and health and healthcare also provide a valuable model for thinking about environmental issues. If we think of the earth as a patient, engineers can be thought of as the earth's doctors. But in a parallel to conventional Western medicine, the earth's doctors focus almost exclusively on curing the symptoms or solving the problems: Rarely is prevention of problems the primary goal. We even use language from healthcare, as in the less than satisfactory "Band-Aid solutions" we are all familiar with.

The Puente Hills landfill in Los Angeles County, for example, is truly a marvel of modern engineering. In the morning you toss out a few tin cans, that old radio and lots of plastic packaging, and within a matter of hours it has been collected, trucked, dumped, spread, packed and buried in an almost seamless, odourless, marvel of engineering and logistics. But what about toxic leachate and air emissions? We don't have a very good handle on how much of this garbage or garbage in any landfill is polluting our air and water. And did anyone ever stop to question the core problem of how and why we produce all of this refuse?

Yes, in fact, someone has, and her name is Annie Leonard, the creator of The Story of Stuff Project. Not only has she thought a great deal about this issue; she has turned it into her life work. I asked Annie to tell me the story of The Story of Stuff. It started off as a way for Annie to show the world how dumb it is to waste as much as we do. She was giving many presentations on the topic before realizing it didn't really make sense for her to fly around America wasting fuel. So she decided to turn her presentation into a wonderful, animated video. It took off like wildfire on YouTube and has been viewed 15 million times to date! Imagine how much jet fuel Annie has saved by not flying to various destinations and presenting a million times! From that simple idea, The Story of Stuff has evolved into a movement of people asking, "Why did we let this happen?" *This* being the creation of the most wasteful society in history.

Annie and I debated whether or not she was focusing on consumers or citizens. She has landed on the idea that if we describe ourselves as consumers, then we're part of the problem—we're self-defining as, in my words, "those who buy stuff." While citizens have a right to work, live and become politically active in a particular country, "consumers" are politically detached, global buyers of goods. Why would we ever want to define ourselves as "consumers"? That was Annie's point. I did push back a little and suggested that we can't just ignore "consumers." Garbage is a colossal problem, and it's getting worse. We saw where all that stuff goes. It becomes garbage containing thousands of toxic chemicals—unknown, untested, untracked and unhealthy. We need to be working on all fronts to stem wasteful production and consumption. And consumers are part of the equation. I agreed with her that at the end of the day, the big issue isn't simply what kinds of stuff we should buy; it's the fact that we need to buy way less stuff, period. Furthermore, that stuff—whether it's a car, a soft drink or a smartphone—needs to be regulated by governments, not by the companies who have no interest apart from endless growth in sales. These regulations need to cover what the products contain and how they are disposed of. Moreover, the costs for proper recycling and disposal need to be built into the products so that scarce natural resources don't end up being buried or burned.

We devote vast amounts of environmental, human and financial resources to digging tiny quantities of metals out of the earth—processing, refining, and manufacturing them into products that are often used for less than a year—sometimes for no more than a 30-second gulp—and then tossing them in a trash heap. Annie told me an amusing story from a Brazilian colleague. He couldn't get over the fact that when presented with a can of pop, "people drink the garbage and throw out the valuable part (the aluminum)." Many countries have started "mining" garbage dumps to find the precious bits of metal that are tossed out with such abandon. I guess the mining engineering challenge of finding needles in haystacks was so much fun the first time around that they figured they'd do it all over. It would have been so much easier if manufacturers, waste managers and governments had had the foresight to set the metals aside before throwing them out with all of the other garbage. The old tossing approach might have been excusable in the 1950s, given what people understood then, but it is downright irresponsible and immoral to continue this practice today.

Of course, it's not as though economists and environmentalists haven't been telling us this since at least the original Earth Day more than 40 years ago. In fact, some of the very first Earth Day placards read "Clean Air, Pure Water." We just can't seem to get our collective heads around the challenges of that simple message because, as much as anything, the rules our society is playing by are the wrong ones. The rules of the game we're playing now are best defined by the Malcolm Forbes maxim: "He who dies with the most toys, wins." We need a different game, with different rules—perhaps "Those who use the least stuff, win." And our economic and regulatory systems need to reinforce that motto with another one—such as this: The more you use, waste, pollute and discard, the more you'll lose financially.

Greening the Economy

Pollution is a broad economic and societal problem, not merely an ecological challenge. Though humans seem to have an inkling that we can't survive without air, water and food, there appears to be less consciousness of the fact that these needs will exist in the future. We're acutely aware of the need for food and water on a daily or weekly basis, but we don't pay as much attention to long-term requirements. Therefore, without deliberate economic instruments designed to capture the long-term environmental damage caused by our consumption (such as fees, tolls, taxes and pollution-trading schemes), we won't succeed in stemming the growth of toxic waste and pollution. This is why we need to "green the economy." Working to ban a chemical that is known to increase the risk of cancer is rewarding and can sometimes be relatively straightforward. But this type of action is not nearly enough. Making the shift to a green economy—one where hazardous chemicals are not manufactured or put in infants' toys in the first place—is more daunting but far more effective. Will we soon see a greener world where our chemistry is green too? Can we overcome the failure of our economy to properly address the health and environmental costs associated with unfettered economic expansion? I'm afraid the jury is still out on those questions. Green chemistry is an exciting concept, but I'm not convinced it will take hold unless we create broader social, cultural and economic change.

Postwar capitalist economies have worked brilliantly in terms of generating wealth and achieving a certain level of economic efficiency. As a 50-year economic experiment, this has fulfilled the forecasts of economists like Milton Friedman and Friedrich Hayek and their free market political followers, Ronald Reagan and Margaret Thatcher. Unfortunately, it has also worked much as economists like Garrett Hardin, Herman Daly, William Baumol and John Kenneth Galbraith predicted. They described various failures of unbridled capitalism leading to the "tragedy of the commons,"[44] "fallacies about growth,"[45] "market failure" due to "externalities"[46] and ultimately the cozy "culture of contentment,"[47] which many of us in developed economies find ourselves enjoying.

I returned to my Teflon pan question, which John Warner said was an excellent example of the complicated challenges facing green chemistry. He has invented a replacement technology for non-stick coatings that is stain resistant, water resistant and doesn't require toxic perfluorinated compounds. It's in the early stages of development, and he requires scarce precommercialization research dollars that our economic system cannot seem to cough up. Without economic incentives or penalties, companies will continue to manufacture toxic products, consumers will buy them thinking they are safe, garbage dumps and incinerators will accept them along with other throwaways containing persistent pollutants, our drinking water will become contaminated and people will get sick. The safer alternatives will never be produced. And so my quest to find out what happens to a Teflon pan led me to discover the failure of global waste systems, plans to retool chemistry education and the need for economic reform.

We need a green economy, one that is more than a contemporary rethinking of sustainable development and that goes far beyond the modest efforts of current Corporate Social Responsibility (CSR) efforts espoused by multinational corporations. The only real answer to getting rid of toxic chemicals is to move from a linear economic model of one-time material extraction, throughput and dumping to a circular economy that mimics Nature by creating closed-loop systems, where resources and nutrients are fed back into the process and the concept of waste is eliminated (see Figure 14).[48] What's more, rather than finding out after the fact that chemicals in our food and cosmetics cause harm, testing protocols need to be followed to ensure that cancer-causing and endocrine-disrupting chemicals are not manufactured in the first place. One such protocol now exists. Developed by a scientific panel of chemical and health experts, including Terry Collins and John Warner, it is called the Tiered Protocol for Endocrine Disruption (TiPED), and it's an example of the new collaborative thinking that may just move us to healthier, greener chemistry.[49] TiPED is a design tool to detect whether or not a chemical will cause endocrine disruption. It consists of five testing tiers, each designed to "broadly interrogate" a chemical's potential effect on the endocrine system of different species.[50]

Figure 14. Linear vs. circular economies

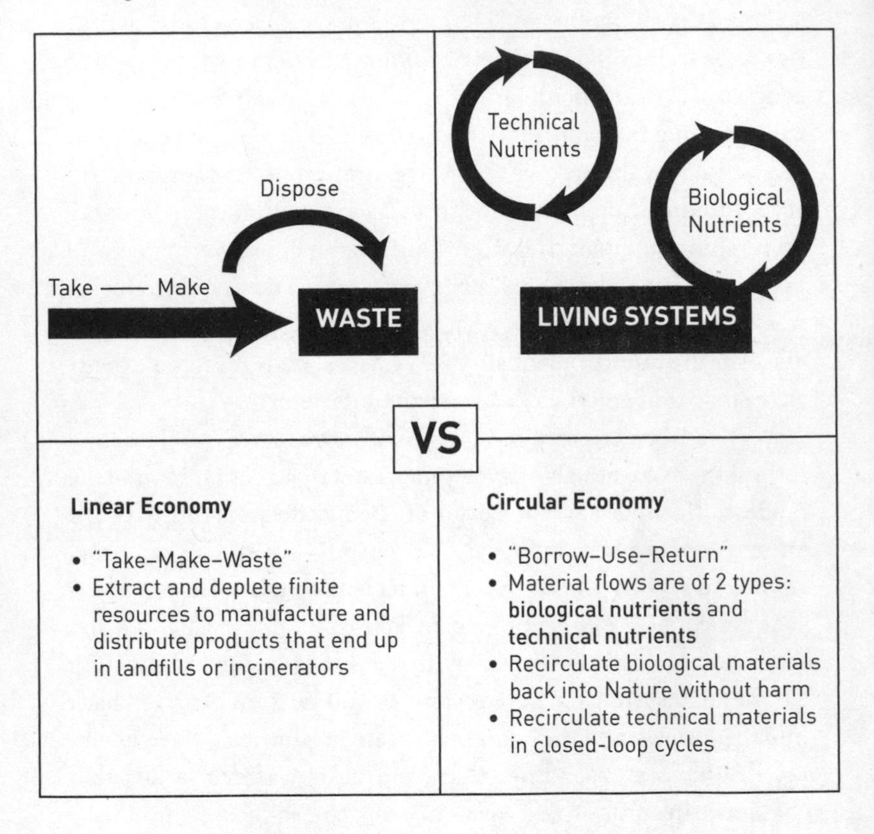

Adapted from the Ellen MacArthur Foundation's Higher Education Resources. © Graham Pritchard / Ellen MacArthur Foundation after W. McDonough and M. Braungart.

There are other hints of a new economy on the horizon as well. Companies are producing a wide array of green products, including cosmetics and cleaning products, as we've described in earlier chapters in this book. And exciting green chemistry breakthroughs are taking place in research labs throughout the globe. At present, German scientists seem to be leading the way—manufacturing biodegradable plastic polymer from milk, using algae or straw to make chemical feedstocks at an economically workable scale and designing colourful pigments based on iron oxide, a.k.a. rust.[51] Even the famous little blue pill, Viagra, has benefited from green chemistry. Chemists at Pfizer, the manufacturer of Viagra, have designed a chemical reaction process that requires dramatically fewer toxic solvents, eliminates the use of tin chloride (an environmental pollutant) and allows for production to happen with only a fraction of the waste involved in the original manufacturing process. [52]

In 2012 global investment in less polluting renewable power generation from sources like wind and solar increased by 17 percent to a record $US 257 billion, and the planet had more than 390 gigawatts of non-hydro renewable energy in place. Tesla Motors is on track to sell over 20,000 fully electric cars in 2013, and plug-in electric vehicle sales in Europe alone are projected to reach nearly 670,000 units annually by 2020.[53] And according to a new report from Pike Research, the total market for energy efficient buildings will reach $103.5 billion by 2017, an increase of more than 50 percent from 2011. So to be sure, there's good news in many industries, but the real transformation has yet to happen.

Monster-sized hills of garbage are still piling up all over the world, and there's a mountain of work to do if we're going to detox our economic systems in time.

SEVEN: THE TOXIN TOXOUT TOP 10

Be careful about reading health books. You may die of a misprint.

—MARK TWAIN

DIET BOOKS ARE DIFFERENT than detox books in one very important respect: Following the advice of the former yields obvious results whereas doing so for the latter may not. If you try a diet and you don't see the pounds coming off, chances are you won't continue. But there's no such built-in quality control with the welter of detox articles and books out there, which provide often conflicting prescriptions for potions and pills and odd foods, often claiming to be "100% GUARANTEED!" to eliminate toxins from your body. Who's to know whether or not these recommendations will really work? It would be nice if you could buy a "Tox-O-Meter" from your local pharmacy, put it under your tongue and get a simple read-out of the various poisons in your body. Alas, this is not possible.

But that's where this book comes in. Through much trial and error, pricey self-testing and numerous uncomfortable situations (Bruce fainting in his basement comes to mind), we have done your homework for you. You can't paralyze yourself into inaction worrying about all the 80,000 or so chemicals that are in commerce today. So we've tried to focus on the ones that exhibit the following two characteristics: (1) The most recent scientific evidence points to them as harmful to human health and (2) Recent science, and our self-tests, shows that you can measurably reduce their levels in your body if you do certain things.

We've summarized our various experimental results in Table 8 in this chapter. As you can see, if you do a few things a little differently, you can dramatically reduce levels of certain pollutants in your body—sometimes very quickly.

This final chapter provides what we hope is a useful summary of the ways you can reduce those toxins (as described in the last six chapters). We've boiled down the voluminous amounts of information in this book, to create a useful "syrup" that you can use to detox your life—10 areas where you can make changes that will benefit your body and reduce the pollutants you and your household are generating.

Perhaps it's because we both do much of our book writing at our respective kitchen tables, but it seems to us the kitchen is really the centre of the modern busy family home. All the household members assemble there many times a day. Any good house party always ends up in the kitchen. And a good number of the choices we're suggesting you make to protect yourself and your family from toxins will determine the kinds of food and other products you'll be storing in your fridge and kitchen cupboards.

Anyway, because most of us spend a lot of time in our kitchens, we boiled the "syrup" down a bit more and reduced it to an easy-to-read list that you can tear out and stick on your refrigerator. It's the Toxin Toxout Top 10 list on page 233 of this book.

Read it, follow its advice . . . and you'll get a whole lot of toxins out of your body in no time flat.

Table 8. Magnitude of changes we achieved through experimentation

Experiment	Changes in Chemical Levels
Chapter 1 *Jessa's and Ray's MEP (phthalate) and methyl paraben levels* phthalates parabens	**MEP** ↓ Levels 10 times on average* when using green products (see Figure 4) **Methyl paraben** ↓ Levels 77 times on average* when using green products (see Figure 4)
Chapter 2 *Levels of pesticide metabolites in nine children* organophosphate (OP) pesticides (dimethyl DAP)	↓ Levels 3 times on average when eating organic (see Figure 5)
Chapter 3 *Levels of heavy metals during chelation* Heavy metals in urine	**Heavy metal** levels ↑ between 4 and 23 times (see Figure 7)
Chapter 4 *Bruce's BPA (bisphenol A) and phthalate (MEP and MBP) levels* BPA and phthalates in sweat	BPA released in sweat all 5 weeks (see Figure 8) MEP and MBP released in sweat all 5 weeks (see Figure 10)

Experiment	Changes in Chemical Levels
Chapter 5 *VOC levels after sitting in new vehicle for 8 hours* VOCs (volatile organic compounds)	**VOC** levels ↑ between 1.2 and 5 times
NBC Experiment** *Andrea Canning's BPA and triclosan levels* BPA (bisphenol A) and triclosan	**BPA** ↓ Levels 88 times **Triclosan** ↓ Levels 99 times

* Increases and decreases were calculated by dividing the peak level by the trough level. The numbers documented here are the average of Ray's and Jessa's respective changes in chemical levels.

** In late 2012 during the writing of this book, we did an experiment with reporter Andrea Canning for *NBC Dateline* that looked at BPA, triclosan and phthalate levels. The experiment, similar to our experiment in Chapter 1 of this book, consisted of three phases: a washout, heavy use of conventional products, followed by another washout where products containing these chemicals were avoided. The results above are from the BPA and triclosan experimentation. The segment aired on *NBC Dateline*, March 24, 2013.

The Most Important "To Do" List of Your Life

Managing what we absorb, breathe, eat and drink is the first line of defence against toxins like phthalates, parabens, pesticides and volatile organic compounds. Once chemicals are in our body, Bruce has some clear ideas on the detox methods that work, and those that are modern-day snake oil. Ultimately, of course, we need to get synthetic toxins out of our economy and our world, but in the meantime, these 10 simple steps (summarized in the Toxin Toxout Top 10 list) will lead to a healthier life for all.

1. Use natural personal-care products that don't contain chemicals like phthalates or parabens.

In a study published just as we were finishing this book, Shanna Swan investigated associations between women's reported use of various types of personal-care products and phthalate metabolite levels in their urine, tested within 24 hours of the women's interviews (see Chapter 1).[1] Swan found that concentrations of MEP (the primary metabolite of diethyl phthalate) in women increased with the number of products used. Swan also found, more generally, that women's more frequent use of these products, particularly perfumes and fragranced products, was associated with higher urinary concentrations of multiple phthalate metabolites.

Looking back at Figure 4 in Chapter 1, you'll see changes in the levels of phthalate (MEP) and methyl paraben in Jessa and Ray, our plucky volunteers for the cosmetics experiment. The graphs in that figure tell us that absorption of the contents of conventional cosmetics and personal-care products clearly contributes to heightened phthalate and paraben levels in our bodies. And they also tell us that using more natural and organic products effectively, and quickly, lowers these levels. Good news, made even better by the fact that natural and organic personal-care products are increasingly easy to find on the market these days. The slow but steady movement of the cosmetics and personal-care industry toward greener products gives consumers much more organic buying power than they've had in the past.

Here are some important tips and tools—and actions to carry out—to reduce toxic chemicals in your personal-care products:

- Less is less. Limit your use of personal-care products whenever possible.

- Use natural and/or organic cosmetics and personal-care products.
- Check out the Environmental Working Group's (EWG's) Skin Deep database for information and safety scores of your products (it includes the scores of natural and organic products).
- Whenever possible, avoid personal-care products with complicated chemical names on the labels (especially products with the label "Fragrance" or "*Parfum*").
- Avoid problem products such as chemical hair straighteners.
- Avoid antibacterial hand soaps and sanitizers, especially the ones that list "Triclosan" on their labels. Instead, wash your hands regularly and vigorously.

2. Eat more organic food to avoid pesticides.

We really are what we eat. In Chapter 2 Rick saw how effective an organic diet is in lowering the levels of pesticides in the bodies of young children. And why should we all avoid those pesticides in the first place? Pesticide exposure has been linked to some very serious negative health effects. Here are just some of them: general developmental problems and cognitive deficits in children, endocrine disruption, non-Hodgkin's lymphoma, low birth weight, reproductive problems, asthma, risk of obesity and diabetes and infertility.[2] In simple terms an organic diet can lower your body pesticide levels. Here's what you can do to reduce the pesticides in your life:

- Eat seasonal local and organic produce whenever you can.
- Choose conventional produce that are lowest in pesticides (usually those with thick skins, such as onions, corn, pineapple and avocados).
- Wash your produce well before eating it.

- Be proud of your chemical-free lawn, yard and neighbourhood.

3. Drink the water from your tap! And lots of it!

We can lower our toxic burden by controlling what we eat, but what about the real "staff of life"—the water that we drink? On average, we are made up of two-thirds water by weight. Add to that the critical role that water plays in our bodily detox mechanisms, and we'd better be sure that we safeguard our drinking water. Water is vital to detoxification—personal and ecological.

Although the huge quantities of toxic waste that we discard can put toxins into our water, public health officials and municipal governments everywhere work together to rigorously test our tap water supplies for hundreds of potential chemical contaminants. And they do it every day. So drinking plenty of tap water, in almost any jurisdiction, will effectively flush toxins out of your system.

But if that doesn't give you enough assurance, you can install an affordable and effective filter in your home. Plenty of them are available. Revisit that handy Table 7 in Chapter 6, comparing some common in-house filters. For many of the chemicals we're concerned about, activated carbon filters are the best bet, and they're affordable. It's a lot easier to install a point-of-use tap filter or to refill that countertop filter than to keep lugging flats of individual-sized water bottles from the grocery store, and it's a whole lot better for the environment and you. So drink up and detox: six glasses of liquid a day for women and nine glasses a day for men (that includes all liquids).

We've also urged you to *stop buying bottled water*. Those bottles, along with countless other plastic items, are a huge part of the landfill problem in the first place, and the plastic bottles themselves may contain chemicals that are best avoided.

4. Use natural fibres and green products like low-VOC paints in your homes and avoid products that might off-gas.

To adapt the wording of the plaque on that colossal beacon of American freedom, the Statue of Liberty: In the modern age, you find the huddled masses indoors, yearning to breathe free. Outdoor atmospheric pollution has long been an obvious area of concern. It's only too easy to notice the car exhaust and the pollutants that power plant smoke stacks, incinerators and manufacturing facilities belch into the air. But with so many of us spending so much time indoors and with so many smelly, off-gassing products surrounding us, indoor air quality has become an area of growing environmental and human health concern. As mentioned in Chapter 5, the average person in an industrialized country now spends over 90 percent of their life indoors, including about 5 percent in enclosed vehicles.[3] With that in mind, air quality inside is more important than it is outside.

To reduce your exposure to off-gassing chemicals in road vehicles, look at the *Healthy Cars* guides produced by Jeff Gearhart and his colleagues at the Ecology Center in Michigan. There's bad news: Toxic chemicals like benzene, phthalates and PBDEs are present in most vehicles, as Rick discovered. But there's also good news: Manufacturers are starting to get rid of them. Check out the report when you're buying a new car to find out who's doing the best job of keeping the toxins out.

You can avoid some toxic chemicals that could off-gas in your car, but what about the ones inside the buildings where most people in industrialized nations are spending the vast majority of their time? As in the case of cars, there's some bad news and some good news. First the bad news: Chemicals like flame retardants and phthalates are getting into the dust that gathers in our homes and offices, and they've been linked to some serious health outcomes like asthma and increased allergenicity.[4] On the upside: You can take preventive steps to reduce your exposure (check out the action items at the end of this section).

But wait! There's even *better* news: The building industry itself is working to create healthier interior environments. Strong voices in this domain are speaking out for net-zero energy, net-zero waste, net-zero water, carbon-neutral and toxin-free buildings. LEED (Leadership in Energy and Environmental Design), North America's pre-eminent green building certification programme, has introduced credits for non-toxic building materials. This is a significant step toward making buildings healthier.

In the meantime, here are some proactive steps you can take to improve indoor air quality and the air you breathe in general. Remember how important it is to breathe in and out deeply—since exhaling is an important detox pathway.

- Open your windows and get outside!
- Incorporate furniture and textiles made from natural fibres into your life, avoid furniture made from polyurethane foam, and reupholster your old furniture whenever it begins to rip.
- Clean and dust interior surfaces frequently (especially those that come in contact with food) and use a vacuum cleaner with a HEPA filter (a type of high-efficiency air filter, usually made from randomly arranged fibreglass—they're designed to capture ultra-fine particles).
- Look for furniture and electronic retailers who carry products that are free of PBDEs.
- Avoid vinyl products.
- Check out the Ecology Center's *Healthy Stuff* new vehicle guide when purchasing a car.
- Use green building materials, like low-VOC paint, when possible.
- When buying or renting a home or office, choose LEED-certified premises when possible.

5. Eat more vegetables and less meat to avoid toxin-grabbing animal fat.

Toxic chemicals are like a bad rash, they keep coming back, and they appear where you least expect or want them. This is the case for toxins like DDT and PCBs, chemicals that were banned as many as 40 years ago. Some of the newer toxic chemicals on the market today—flame retardants like PBDEs and commonly used pesticides—also bioaccumulate. These toxins, transported by air and water, are lipophilic (fat-loving), so they find their way into the fat cells of wildlife like fish and move up the food chain into humans. A bummer for both the salmon and the salmon eater.

What's the solution to avoiding these persistently pesky chemicals in the food chain? Well, it's not all that different from what doctors and nutritionists (and many mothers) already recommend: Eat your veggies and stay away from fatty foods!

And the research supports this advice—from the point of view of both detoxification and general health. A 2010 Korean study found that changing to a vegetarian diet (i.e., reduced consumption of animal fats) reduced levels of environmental chemicals like phthalates and antibiotics after just five days.[5] More evidence that substantiates the claim that a plant-based diet improves health has arisen from a dark time and place: Nazi-occupied Norway in World War II. After occupation the Nazis confiscated all local livestock to provide supplies for their own troops. As a result, food like meat, eggs and dairy were heavily rationed among the Norwegian population, and up went their consumption of veggies and plant-based supplements.[6] A later study in *The Lancet* looked at mortality rates from heart disease between 1927 and 1948 and found a significant decrease in these rates during the time of occupation and the forced vegetarian diets.[7] Strong proof of the value of a vegetable-rich diet despite the terrible circumstances.

6. Sweat more—toxic chemicals like BPA and phthalates leave your body through your sweat.

When researching his chapters, Bruce learned firsthand about the sheer volume of resources (useful or otherwise) that exist on the topic of diet- and exercise-based detox routines. Do check out some of the resources we recommend at the end of the book. Saunas have been used for spiritual and therapeutic reasons alike, across cultures, for hundreds and hundreds of years. The sauna detox experiment Bruce undertook was a novel design. And though it wasn't possible to show before and after body concentrations, the experiment demonstrated unmistakably that synthetic chemicals like BPA are removed through sweating—one of our body's most basic natural detox methods.

7. Exercise!

Though the science of detox is in its infancy, some methods do clearly work—as evidenced by the research of Stephen Genuis.[8] But some methods don't work. The effectiveness of a given detox method often depends on the toxic chemical in question. Some of the common ones that we've mentioned in this book (phthalates and parabens, for instance) are metabolized and excreted quickly through the body's natural mechanisms, whereas others find their way into fat cells and therefore bioaccumulate. To help get rid of these "lipophilic" (fat-loving) chemicals, do plenty of exercise. This breaks down the fat cells, releasing the stored toxins and allowing them to be excreted via lung exhalation and sweat.[9]

In one recent example, a study carried out at the University of Montreal evaluated the impact of physical exertion on human exposure to the volatile organic compounds (VOCs) toluene and n-hexane. The researchers found that individuals subjected to toluene and n-hexane in equal amounts produced significantly higher alveolar air toluene and n-hexane biomarker levels when undergoing increased exertion as compared to resting levels.[10] Well, great. But what exactly does that mean? Essentially, with increased ventilation (i.e., breathing) from physical exercise, higher concentrations of toluene and n-hexane levels were being exhaled from the lungs.[11]

8. Avoid wacky quick-fix detoxes and optimize your body's natural detox mechanisms by adopting a detox lifestyle.

Since we know that chemicals are entering our bodies on a daily basis, shouldn't we make sure that they're leaving our bodies on a daily basis? And shouldn't we train our bodies to be in the best possible condition to do this? Ridding your body of pollutants you can't actively avoid requires a certain lifestyle approach that, frankly, may take a bit of effort. Recall that myth-busting Table 6 from Chapter 3 about the effectiveness of various detox treatments? As Bruce explored the depths of the detox industry, he debunked some of the phony detox techniques that are being peddled on the market today. He also learned that to succeed in eliminating toxic chemicals, we have to overthrow the notion of quick-fix, fad-based diets and cleanses.

One study, for instance, examined the use of ionic footbaths and their ability to remove potential toxic elements (like arsenic and lead) from the body. Samples of water were taken before and after the footbath treatment sessions, and it was discovered that no significant amount of toxic elements had been released.[12] Further, the researchers found the highest concentrations of metals in the water after a footbath were those associated with the metal array used to conduct the current in the water. Unfortunately, the purveyors of footbaths appear to care so little about your individual health, they're trying to trick you into using their product. Focus instead on our healthy detox lifestyle recommendations.

9. Buy less, buy green.

As consumers we need to protect ourselves and our families by making informed choices—since it appears that many corporations have little concern for our health. Fortunately, however, there's also a growing trend among small and big businesses to do better by us. We've already talked about the myriad companies that are "detoxing" their products—producing safer makeup, personal-care products, cleaning supplies and house paint, for instance. And these items are also easier to find because of market demand.

A great illustration of this point: As this book was being written, Johnson & Johnson voluntarily phased out the use of formaldehyde-releasing preservatives in their baby shampoo sold in Canada, the U.S., China, Australia and Indonesia. In September 2013, in quick succession Procter & Gamble announced that phthalates and triclosan would be eliminated from its products by 2014, and Walmart said it will require its suppliers to phase out 10 hazardous chemicals from personal care products, cosmetics and cleaning products sold in its stores. All this was a direct response to calls for action from consumers (many of them readers of *Slow Death by Rubber Duck*) and groups like the Campaign for Safe Cosmetics and Environmental Defence Canada.

Things have changed. While doing some research for a media story in New York, we popped into a local grocery store to purchase some BPA microwaveable food containers. Much to our amazement (and dismay, in terms of the experiment we were preparing), we couldn't find any. All the brand-name plastic containers were now BPA free. This is a clear signal that despite the whining from chemical companies, the days are numbered for many synthetic toxins thanks to active and vocal consumers, like you.

These may appear to be small steps, but they are a big reminder that consumers have the power to demand healthier products and win.

As we, individually, become more adept at the Toxin Toxout

Top 10 list, we must also recognize that these same steps need to be applied to modern society as a whole: We have to get our collective act together to create a greener, less toxic economy.

In our role as consumers, we're doing a pretty good job of just that—consuming. Think about Bruce's trash travels and tales: Individuals and society as a whole are creating more waste per day than ever before. And that's *after* we take into account the so-called success of the recycling and waste diversion programmes that towns, cities, provinces and states have developed in response to the growing trash problem. The waste piles growing throughout the world are a reflection of our society's ability to consume, consume and then discard.

Garbage itself is not so much the problem as the fact that, too often, much of it is toxic, and we make and consume way too much. People like Terry Collins, John Warner and Janine Benyus are leaders in green chemistry, and they're working to make materials and designs available for the stuff we already create. Software programmes are being developed to better predict the toxicity of chemicals so we can avoid using any that will harm human health, and chemists are looking into shifting away from petroleum-based chemical manufacturing. Petroleum is a common thread running through our toxic world—from perfumes to pesticides to plastics. In the world of biomimicry, scientists are taking a step back and looking to Nature for inspiration, where they would have previously resorted to chemical engineering solutions. But does this mean that we're just learning to make more stuff in a less harmful way? Green chemistry may be gaining ground, but it solves only part of the problem.

10. Support politicians who believe in a greener economy and organizations that work for a cleaner environment.

Though much of this book has focused on how to exercise our power as consumers by making different choices, only by mobilizing our power as consumers *and also as citizens* will the prospect of a modern, green economy emerge. If we think of ourselves only as consumers, we'll end up as part of the problem. We don't need more consumers; we need active citizens, as Annie Leonard reminded us.

To meet the challenges of detoxing our economy, citizens will need to be informed and engaged. We also need to produce governments that will put public health ahead of large corporate profits. Governments decide which chemicals can be sold or not, where our garbage ends up and what levels of toxins will be allowed in our food, water and air. That's why we need to elect politicians who understand the importance of keeping toxic chemicals out of our lives. You can help create a green economy by electing politicians who support incentives and disincentives for consumers and manufacturers that will lead to the production of greener products and less waste. We all respond to economic signals, and we need governments to develop effective pricing policies that capture the full cost of pollution, including the pollution of our bodies by toxic chemicals.

Does it make a difference to elect the right politicians? Darn right it does—and in a measurable way. A few years ago the Ontario government committed to banning certain lawn pesticides. The main Opposition party didn't agree. Guess what? When the government was re-elected, Ontario banned those pesticides and lo and behold, levels of common herbicides like 2,4-D and MCPP were down as much as 94 percent in the urban streams tested just one year after the ban came into effect.[13] Those streams replenish the primary source of drinking water for millions of Ontarians, and were made cleaner because a majority of Members of Provincial Parliament caused it to happen. Like the major and positive corporate announcements that we mentioned a couple of pages ago, a big reason this pesticide decision occurred is because of the pushing of environmental and citizen organizations, in this case the Canadian Association of Physicians for the Environment.

Start Somewhere

Is all of the above a perfect recipe for being toxin free? Of course not. But by following these guidelines, you'll make a huge difference in creating a cleaner, healthier lifestyle. It's the most reliable, evidence-based prescription that we know of to get chemicals out of your environment and your body. Scientists have observed that some of us are more susceptible to the effects of toxic chemicals than others. But how do we know? Well, this issue is similar to the smoking debate. Rick's beloved grandmother Marjorie Braive could have been the poster girl for the tobacco industry. She smoked at least a pack a day of Du Maurier Regulars dating back to the 1930s. In the early 1970s she had a mastectomy but then carried on, undeterred, to the ripe old age of 85. One of Rick's enduring childhood memories is of her swimming off the dock of their Adirondack cottage, a floppy-brimmed straw sunhat on her head and a cigarette hanging out of her mouth. Her family sprinkled a few cigarettes on top of her coffin to carry her through to the afterlife.

On the other hand, how many of us know of people whose lives were clearly cut short by the effects of smoking? Rick's grandfather on his father's side, a heavy smoker, dropped dead of a massive heart attack in his mid-40s.

The effects of pollution on various individuals is similar. Some of us have constitutions that are resilient enough to remain relatively unaffected by toxic synthetic chemicals. Others can be damaged by even apparently minor exposure. All of us benefit when exposure to cigarette smoke is reduced and the same is true of exposure to synthetic chemicals.

The key is to start. Start somewhere.

Now rip out page 233, stick it on your refrigerator and get going!

TOXIN TOXOUT TOP 10

1. Use natural personal-care products that don't contain chemicals like phthalates or parabens.
2. Eat more organic food to avoid pesticides.
3. Drink the water from your tap! And lots of it!
4. Use natural fibres and green products like low-VOC paints in your home and avoid products that might off-gas.
5. Eat more vegetables and less meat to avoid toxin-grabbing animal fat.
6. Sweat more—toxic chemicals like BPA and phthalates leave your body through your sweat.
7. Exercise!
8. Avoid wacky quick-fix detoxes and optimize your body's natural detox mechanisms by adopting a detox lifestyle.
9. Buy less, buy green.
10. Support politicians who believe in a greener economy and organizations that work for a cleaner environment.

www.toxintoxout.com

RESOURCE GUIDE/FURTHER READING

Groups to Access Online

Breast Cancer Fund *www.breastcancerfund.org*
Broadbent Institute *www.broadbentinstitute.ca*
Campaign for Safe Cosmetics *www.safecosmetics.org*
Canadian Association of Physicians for the Environment *www.cape.ca*
Canadian Breast Cancer Foundation *www.cbcf.org*
Canadian Organic Growers *www.cog.ca*
Center for Environmental Health *www.ceh.org*
Ecology Center *www.ecologycenter.org*
Environmental Defence Canada *www.environmentaldefence.ca*
Environmental Working Group *www.ewg.org*
Natural Resources Defense Council *www.nrdc.org*
Organic Center *www.organic-center.org*
Organic Monitor *www.organicmonitor.com*
Organic Trade Association *www.ota.com*

Databases and Other Online Resources

- Canada. Government of Canada's Chemicals Management Plan. Everything you wanted to know about the regulation of synthetic chemicals in Canada. *http://www.chemicalsubstanceschimiques.gc.ca/plan/index-eng.php.*
- *Ecology Center Healthy Car Report*. See how your car checks

out, and use this report as a guide when purchasing new cars: *http://www.healthystuff.org/get-stuff.php?group-report=Cars.*

- Environmental Defence. Download the pocket shopping guide for an easy-to-use handbook of chemicals to avoid: *http://environmentaldefence.ca/reports/just-beautiful-personal-care-products-pocket-shopping-guide.*
- Environmental Health News Service. Keep up to date on environmental health (and more) with this daily listing of media clippings covering many issues, as well as updates on scientific research: *www.environmentalhealthnews.org.*
- *Environmental Working Group Food News*. Check out their Clean 15 and Dirty Dozen Guides: *http://www.ewg.org/foodnews/.*
- Environmental Working Group Skin Deep Cosmetics Database. Search for your favourite products here to see how they rate and check out the database of the safest cosmetics and personal-care products: *http://www.ewg.org/skindeep/.*
- NSF International Product and Service Listings Database. Gives product and certification information about various types of drinking water treatment units *http://www.nsf.org/Certified/dwtu/.*

Further Reading

Anastas, Paul, and John Warner. *Green Chemistry: Theory and Practice*. Oxford: Oxford University Press, 2000.

Baker, Nena. *The Body Toxic: How the Hazardous Chemistry of Everyday Things Threatens Our Health and Well-Being*. Portland, OR: North Point Press, 2008.

Benyus, Janine M. *Biomimicry: Innovation Inspired by Nature*. New York: William Morrow, 1997.

Carson, Rachel. *Silent Spring*. New York: Mariner Books, 1962.

Colborn, Theo, Dianne Dumanoski and J.P. Myers. *Our Stolen Future*. New York: Dutton/Penguin, 2002.
Crinnion, Walter. *Clean, Green & Lean: Get Rid of the Toxins That Make You Fat*. Hoboken, NJ: John Wiley & Sons, 2010.
Davis, Devra. *The Secret History of the War on Cancer*. New York: Basic Books, 2007.
Duke, Deanna. *The Non-Toxic Avenger: What You Don't Know Can Hurt You*. Gabriola Island, BC: New Society Publishers, 2011.
Elton, Sarah. *Locavore: From Farmers' Fields to Rooftop Gardens—How Canadians Are Changing the Way We Eat*. Toronto: HarperCollins Canada, 2010.
Freinkel, Susan. *Plastic: A Toxic Love Story*. New York: Houghton Mifflin Harcourt, 2011.
Humes, Edward. *Garbology: Our Dirty Love Affair with Trash*. New York: The Penguin Group, 2012.
Macauley, David. *Elemental Philosophy: Earth, Air, Fire and Water as Environmental Ideas*. New York: State University of New York Press, 2010.
Malkan, Stacy. *Not Just a Pretty Face: The Ugly Side of the Beauty Industry*. Gabriola Island, BC: New Society Publishers, 2007.
McLennan, Jason F. *The Philosophy of Sustainable Design: The Future of Architecture*. Bainbridge Island, WA: Ecotone Publishing, 2004.
Pollan, Michael. *Food Rules: An Eater's Manual*. New York: Penguin Books, 2009.
Vasil, Adria. *Ecoholic: Your Guide to the Most Environmentally Friendly Information, Products and Services in Canada*. Toronto: Vintage Canada, 2007.
——. *Ecoholic Body: Your Ultimate Earth-Friendly Guide to Living Healthy and Looking Good*. Toronto: Vintage Canada, 2012.
——. *Ecoholic Home: The Greenest, Cleanest and Most Energy-Efficient Information under One (Canadian) Roof*. Toronto: Vintage Canada, 2009.

Vogel, Sarah. *Is It Safe? BPA and the Struggle to Define the Safety of Chemicals*. Berkeley: University of California Press, 2012.

Watson, Brenda. *The Detox Strategy: Vibrant Health in Five Easy Steps*. New York: Free Press, 2008.

Williams, Florence. *Breasts: A Natural and Unnatural History*. New York: W.W. Norton, 2012.

Wylde, Bryce. *Wylde on Health: Your Best Choices in the World of Natural Health.* Toronto: Random House Canada, 2012.

NOTES

Introduction

1 Environmental Working Group, "Cord Blood Contaminants in Minority Newborns," 2009, accessed December 3, 2012, http://www.ewg.org/research/minority-cord-blood-report/executive-summary. Also check out EWG's Human Toxome Project for the latest on their research into pollution in people. Accessed June 17, 2013, http://www.ewg.org/sites/humantoxome/.

2 Environmental Defence Canada, *Pre-polluted: The first report on toxic substances in the umbilical cord blood of Canadian newborns*, Toronto, June 2013.

3 United Nations Environment Program, "Global Chemicals Outlook: Towards Sound Management of Chemicals," 2012, accessed December 3, 2012, http://www.unep.org/pdf/GCO_Synthesis%20Report_CBDTIE_UNEP_September5_2012.pdf.

4 S. Sorian, P. Alonso-Magdalena, M. Garcia-Arevalo, A. Novials, S. Muhammed, A. Salehi, J. Gustafsson, I. Queseda and A. Nadal, "Rapid Insulinotropic Action of Low Doses of Bisphenol-A on Mouse and Human Islets of Langerhans: Role of Estrogen Receptor ß," *PLoS ONE* 7 (2012), no. 2; D. Melzer, N. Rice, C. Lewis, W. Henley and T. Galloway, "Association of Urinary Bisphenol A Concentration with Heart Disease: Evidence from NHANES 2003/06," *PLoS One* 5 (2010), no. 1; S. Ehrlich, P. Williams, S. Missmer, J. Flaws, K. Berry, A. Calafat, X. Ye, J. Petrozza, D. Wright and H. Hauser, "Urinary Bisphenol A Concentrations and Implantation Failure among Women Undergoing *in Vitro* Fertilization," *Environmental Health Perspectives*, 120 (2012): 978–83.

5 Take a look at the following articles: United States Department of Health and Human Services, Centers for Disease Control and Prevention and National Center for Environmental Health, *Fourth National Report on Human Exposure to Environmental Chemicals* (Atlanta, GA: United States Department of Health and Human Services, 2009); E. Rees Clayton, M. Todd, J. Dowd and A. Aiello, "The Impact of Bisphenol A and Triclosan on Immune Parameters in the U.S. Population, NHANES 2003–2006," *Environmental Health Perspectives* 199 (2010): 390–96; "Scientists Discover That Antimicrobial Wipes and Soaps May Be Making You (and Society) Sick," *Scientific American*, July 2011, accessed February 12, 2012, http://blogs.scientificamerican.com/guest-blog/2011/07/05/scientists-discover-that-antimicrobial-wipes-and-soaps-may-be-making-you-and-society-sick/.

6 C. Gallagher and J. Meliker, "Mercury and Thyroid Autoantibodies in U.S. Women, NHANES 2007–2008," *Environment International* 40 (2012): 39–43. See also B. Gump, J. MacKenzie, A. Dumas, C. Palmer, P. Parsons, Z. Segu, Y. Mechref and K. Bendinskas, "Fish Consumption, Low-Level Mercury, Lipids and Inflammatory Markers in Children," *Environmental Research* 112 (2012): 204–11.

7 C8 Science Panel official website, accessed July 9, 2013, http://www.c8sciencepanel.org.

8 For example, see I. Andersen, O. Voie, F. Fonnum and E. Mariussen, "Effects of Methyl Mercury in Combination with Polychlorinated Biphenyls and Brominated Flame Retardants on the Uptake of Glutamate in Rat Brain Synaptosomes: A Mathematical Approach for the Study of Mixtures," *Toxicological Sciences* 112 (2009): 175–84.

9 Check out J. Whitfield, V. Dy, R. McQuilty, G. Zhu, A. Heath, G. Montgomery and N. Martin, "Genetic Effects on Toxic and Essential Elements in Humans: Arsenic, Cadmium, Copper, Lead, Mercury, Selenium and Zinc in Erythrocytes," *Environmental Health Perspectives* 188 (2010): 776-82. Also check out H. Mortensen and S. Euling, "Integrating Mechanistic and Polymorphism Data to Characterize Human Genetic Susceptibility for Environmental Chemical Risk

Assessment in the 21st Century," *Toxicology and Applied Pharmacology* (2011), pii: S0041-008X(11)00025-1. doi: 10.1016/j.taap.2011.01.015 [Epub ahead of print].

10 IFOAM and FiBL, *The World of Organic Agriculture: Statistics and Emerging Trends 2012* (February 2012).

11 Agriculture and Agri-Food Canada, *Market Trends: Organics—Market Analysis Report, 2010*, accessed December 3, 2012, http://www.ats.agr.gc.ca/inter/5619-eng.htm#b.

12 Trend Watching, "12 Crucial Consumer Trends for 2012," accessed December 3, 2012, http://trendwatching.com/trends/12trends2012/?ecocycology.

13 "Sustainability Nears a Tipping Point," *MIT Sloan Management Review* 53 (Winter 2012): 69–74.

14 Ibid.

15 L. Tan, N. Neilsen, D. Young and Z. Trizna, "Use of Antimicrobial Agents in Consumer Products," *Archives of Dermatology* 138 (2002): 1082–86; Environmental Defence Canada, *The Trouble with Triclosan*, Toronto, November 2012; Canada, Health Canada, "Canada Concludes Preliminary Assessment of Triclosan," press release, March 30, 2012, accessed December 3, 2012, http://www.hc-sc.gc.ca/ahc-asc/media/nr-cp/_2012/2012-48-eng.php.

16 United States, National Resources Defense Council, "Health Issues," accessed February 12, 2012, http://www.nrdc.org/breastmilk/ddt.asp.

17 D. Smith, "Worldwide Trends in DDT Levels in Human Milk," *International Journal of Epidemiology* 28 (1999): 179–88.

18 POP Technical Note, accessed December 3, 2012, http://www.who.int/foodsafety/chem/pops/en/.

19 P. Pinsky and M. Lorber, "A Model to Evaluate Past Exposure to 2,3,7,8-TCDD," *Journal of Exposure Analysis and Environmental Epidemiology* 8 (1998): 187–206.

20 Canada, Health Canada, "Final Human Health State of the Science Report on Lead," February 2013, accessed April 19, 2013, http://www.hc-sc.gc.ca/ewh-semt/pubs/contaminants/dhhssrl-rpecscepsh/index-eng.php.

Chapter 1

1 EcoFocus Trend Survey 2010–2012, accessed July 30, 2012, http://ecofocusworldwide.com/?p=776.

2 Personal Care Products Council, "About Us," accessed August 1, 2012, http://www.personalcarecouncil.org/about-us/about-personal-care-products-council,

3 Estée Lauder Official Website, accessed July 30, 2012, www.esteelauder.com.

4 United States Department of Labor, Occupational Safety and Health Administration, "Hazard Alert Update: Hair Smoothing Products That Could Release Formaldehyde," accessed August 1, 2012, http://www.osha.gov/SLTC/formaldehyde/hazard_alert.html.

5 California, State of California Department of Justice, Office of the Attorney General, "Attorney General Kamala D. Harris Announces Settlement Requiring Honest Advertising over Brazilian Blowout Products," press release, January 30, 2012, accessed August 1, 2012, http://oag.ca.gov/news/press-releases/attorney-general-kamala-d-harris-announces-settlement-requiring-honest.

6 Canada, Health Canada, "Media Advisory and Product Safety Recall: Brazilian Blowout Solution Contains Formaldehyde," October 26, 2010, accessed August 2, 2012, http://www.healthycanadians.gc.ca/recall-alert-rappel-avis/hc-sc/2010/13437a-eng.php.

7 Environmental Defence Canada, *Not So Sexy: The Health Risks of Secret Chemicals in Fragrance* (Toronto: Environmental Defence, 2010).

8 The word "girlcott" was first coined in 2005 by a group of high school girls who were protesting sexist T-shirt slogans by Abercrombie & Fitch. In her book *Not Just a Pretty Face* (Gabriola

Island, BC: New Society Publishers, 2007), Stacy Malkan discusses synthetic chemicals and quotes Dr. Devra Davis as saying: "Boycotts mean saying no. Girlcotts mean yes. Women are the main purchasers of products and take responsibility for what goes into the home. We can organize to change market forces by saying we don't want cancer-causing products, and we do want safer products. When enough women get together, we can make things happen."

9 C. Martina, B. Weiss and S. Swann, "Lifestyle Behaviours Associated with Exposures to Endocrine Disruptors," *Neurotoxicology* 6 (2012): 1247–1433, http://dx.doi.org/10.1016/j.neuro.2012.05.016.

10 O. Geiss, S. Tirendi, J. Barrero-Moreno and D. Kotzias, "Investigation of Volatile Organic Compounds and Phthalates Present in the Cabin Air of Used Private Cars," *Environment International* 35 (2009): 1188–95.

11 R. Rudel, J. Gray, C. Engel, T. Rawsthorne, R. Dodson, J. Ackerman, J. Rizzo, J. Nudelman and J. Brody, "Food Packaging and Bisphenol A and Bis(2-ethyhexyl) Phthalate Exposure: Findings from a Dietary Intervention," *Environmental Health Perspectives* 119 (2011): 914–20.

12 R. Kwapniewski, S. Kozaczka, R. Hauser, M. Silva, A. Calafat and D. Duty, "Occupational Exposure to Dibutyl Phthalate among Manicurists," *Journal of Occupational and Environmental Medicine* 50 (2008): 705–11.

13 L. Parlett, A. Calafat and S. Swan, "Women's Exposure to Phthalates in Relation to Use of Personal Care Products," *Journal of Exposure Science and Environmental Epidemiology*, November 21, 2012, doi: 10.1038/jes.2012.105.

14 J. Meeker, S. Sathyanarayana and S. Swan, "Phthalates and Other Additives in Plastics: Human Exposure and Associated Health Outcomes," *Philosophical Transactions of the Royal Society of London B: Biological Sciences* 364 (2009): 2097–113.

15 S. Swan, K. Main, F. Liu, S. Stewart, R. Kruse, A. Calafat, C. Mao, J. Redmon, C. Ternand, S. Sullivan and J. Teague, "Decrease in Anogenital Distance among Male Infants with Prenatal Phthalate Exposure," *Environmental Health Perspectives* 113 (2005): 1056–61.

16 S. Swan, F. Liu, M. Hines, R. Kruse, C. Wang, J. Redmon, A. Sparks and B.Weiss, "Prenatal Phthalate Exposure and Reduced Masculine Play in Boys," *International Journal of Andrology* 33 (2010): 259–69.

17 A. Calafat, X. Ye, L. Wong, A. Bishop and L. Needham, "Urinary Concentrations of Four Parabens in the U.S. Population: NHANES 2005–2006," *Environmental Health Perspectives* 118 (2010): 679–85.

18 P. Darbre, A. Aljarrah, W. Miller, N. Coldham, M. Sauer and G. Pope, "Concentrations of Parabens in Human Breast Tumours," *Journal of Applied Toxicology* 24 (2004): 5–13.

19 N. Janjua, G. Mortensen, A. Andersson, B. Kongshoj, N. Skakkebaek and H. Wulf, "Systemic Uptake of Diethyl Phthalate, Dibutyl Phthalate and Butyl Paraben Following Whole-Body Topical Application and Reproductive and Thyroid Hormone Levels in Humans," *Environmental Science and Technology* 41 (2007): 5564–70.

20 L. Barr, G. Metaxas, C. Harbach, L. Savoy and P. Darbre, "Measurement of Paraben Concentrations in Human Breast Tissue at Serial Locations across the Breast from Axilla to Sternum," *Journal of Applied Toxicology* 3 (2012): 219–32.

21 P. Darbre, D. Pugazhendhi and F. Mannello, "Aluminium and Human Breast Diseases," *Journal of Inorganic Biochemistry* 105 (2011): 1484–88.

22 CosmeticsInfo.Org., "Paraben Information," accessed August 10, 2012, http://www.cosmeticsinfo.org/HBI/9.

23 S. Khanna and P. Darbre, "Parabens Enable Suspension Growth of MCF-10A Immortalized, Non-transformed Human Breast Epithelial Cells," *Journal of Applied Toxicology*, June 29, 2012, doi: 10.1002/jat.2753 [Epub ahead of print].

24 P. Darbre and A. Charles, "Environmental Oestrogens and Breast Cancer: Evidence for Combined Involvement of Dietary, Household and Cosmetic Xenoestrogens,"*Anticancer Research* 30 (2010): 815–28.

25 In *Slow Death by Rubber Duck* we conducted a much simpler experiment comparing phthalate levels during and after wearing conventional

personal-care products. In this experiment we wanted to design something that would be more useful for consumers: a comparison of synthetic chemical levels from conventional and natural cosmetics, since virtually nobody will go cosmetics free. As far as we know, the paraben experiment and comparison of levels resulting from wearing conventional and natural cosmetics is a "first."

26 Following are the products lists for Jessa Blades and Ray Civello:

Jessa Blades—Conventional Products

Neutrogena Pore Refining Cleanser

Neutrogena Alcohol-Free Toner

Estée Lauder DayWear Plus Multi Protection Anti-Oxidant Creme

Pantene Pro-V Moisture Renewal 2-in-1 Shampoo

Pantene Pro-V Moisture Renewal Conditioner

Herbal Essences Set Me Up Mousse and Herbal Essences Set Me Up Extra Hold Styling Gel

John Frieda Luxurious Volume Extra Hold Hairspray

Olay Body Ultra Moisture Body Wash with Shea Butter

Irish Spring Original Deodorant Soap

Lady Speed Stick 24/7 Antiperspirant Deodorant—Cool Breeze

Vaseline Intensive Care Total Moisture Dry Skin Lotion

Alfred Sung Forever perfume

Dial Complete Antibacterial Foaming Hand Wash

Covergirl TruBlend Pressed Powder

Physicians Formula Summer Eclipse Radiant Bronzing Powder

Revlon ColorStay Mineral Blush

L'Oréal Neutrals eyeshadow

Covergirl Professional Super Thick Lash Mascara

Revlon ColorStay Overtime Lipcolor

Sally Hansen Insta-dri fast dry nail color

Jessa Blades—Green Products

Aubrey Organics shampoo

Aubrey Organics conditioner

Dr. Hauschka eye shadow

Zosimos botanicals eyeliner

Couleur Caramel mascara

Organic Pharmacy honey and jasmine mask

Earth Tu Face Body Butter

TMS Beauty concealer

Jane Iredale PurePressed powder

Nine Naturals body wash

Sprout lip balm

Organic Pharmacy blush

Ray Civello—Conventional Products

Gillette Fusion shaving gel

Neutrogena Men Post Shave Lotion

Neutrogena Pore Refining Cleanser

Nivea for Men moisturizer

Pantene Pro-V Always Smooth Shampoo

Pantene Pro-V Sheer Volume Conditioner

Bed Head for Men Matte Separation Workable Wax

Dove Men+Care Deep Clean bodywash

Irish Spring Original Deodorant Soap

Old Spice deodorant

Vaseline Men Fast Absorbing Lotion

Axe body spray

Dial Complete Antibacterial Foaming Hand Wash

Ray Civello—Green Products

Aveda Caribbean Therapy Body Creme

Aveda Calming Body Cleanser

Aveda Rosemary Mint Hand and Body Wash

Aveda Hand Relief

Aveda Foot Relief

Aveda Tourmaline Charged Hydrating Creme

Aveda Men Pure-Formance Shampoo

Aveda Men Pure-Formance Conditioner

Aveda Men Pure-Formance Grooming Clay

Aveda Men Pure-Formance Shave Cream

Aveda Men Pure-Formance Dual Action Aftershave

Though it's impossible to nail down with complete precision which amount of phthalates and parabens came from which specific products, the selection—according to a variety of sources—both likely contained the chemicals in question and replicated the typical product selection of countless consumers.

27 R. Dodson, M. Nishioka, L. Standley, L. Perovich, J. Brody and R. Rudel, "Endocrine Disruptors and Asthma-Associated Chemicals in Consumer Products," *Environmental Health Perspectives* 20 (2012): 935–43.

28 Environmental Working Group, "Skin Deep Cosmetics Database," accessed September 20, 2012, http://www.ewg.org/skindeep/.

29 The lab in B.C. analyzed the urine samples for the following phthalates and parabens: monoethyl phthalate (MEP), mono(2-ethylhexyl) phthalate (MEHP), mono(2-ethyl-5-hydroxyhexyl) phthalate (MEHHP), mono(2-ethyl-5-oxohexyl) phthalate (MEOHP), mono-benzyl phthalate (MBzP), mono(3-carboxypropyl) phthalate (MCPP),

monomethyl phthalate (MMP), mono-isobutyl phthalate (MiBP), and mono-n-butyl phthalate (MnBP) and methyl, ethyl, *n*-propyl, butyl, and benzyl parabens.

30 N. Janjua, H. Frederiksen, N. Skakkebaek, H. Wulf and A. Andersson, "Urinary Excretion of Phthalates and Parabens after Repeated Whole-Body Application in Humans," *International Journal of Andrology* 3 (2008): 118–30.

31 Tonnages of shea nuts: Joseph Funt, Managing Director, Global Shea Alliance, personal correspondence to author, July 18, 2012.

32 T. Schettler, "Human Exposure to Phthalates via Consumer Products," *International Journal of Andrology* 29 (2006): 134–39.

33 M. Wormuth, M. Scheringer, M. Vollenweider and K. Hungerbühler, "What Are the Sources of Exposure to Eight Frequently Used Phthalic Acid Esters in Europeans?" *Risk Analysis* 26: 803–24.

34 C.G. Bornehag, J. Sundell, C. Weschler, T. Sigsgaard, B. Lundgren, M. Hasselgren and L. Hägerhed-Engman, "The Association between Asthma and Allergic Symptoms in Children and Phthalates in House Dust: A Nested Case-Control Study," *Environmental Health Perspectives* 112 (2004): 1393–97; C. Bornehag and E. Nanberg, "Phthalate Exposure and Asthma in Children," *International Journal of Andrology* 33 (2009): 1-13; M. Larsson, L. Hägerhed-Engman, B. Kolarik, P. James, F. Lundin, S. Janson, J. Sundell and C.G. Bornehag, "PVC—as Flooring Material—and Its Association with Incident Asthma in a Swedish Child Cohort Study," *Indoor Air* 20 (2010): 494–501; N. Hsu, C.G. Bornehag, J. Sundell, C. Lee, J. Wang, H. Chang, C. Chen, P. Wu and H. Su, "Predicted Risk of Childhood Allergy, Asthma and Reported Symptoms Using Measured Phthalate Exposure in Dust and Urine,"*Indoor Air* 22 (2012): 186–99.

35 G. Toft, B. Jönsson, C. Lindh, T. Kold Jensen, N. Hjollund, A. Vested and J. Bonde, "Association between Pregnancy Loss and Urinary Phthalate Levels around the Time of Conception," *Environmental Health Perspectives* 120 (2012), no. 2: 458–63.

36 S. Teitelbaum, A. Calafat, N. Mervish, E. Moshier, N. Vangeepuram, M. Galvez, M. Silva, B. Brenner and M. Wolff, "Associations between Phthalate Metabolite Urinary Concentrations and Body Size Measures in New York City Children," *Environmental Research* 112 (2012): 186–93.

37 Swan et al., "Decrease in Anogenital Distance," 1056–61; K. Main, C. Mortensen, M. Kaleva, K. Boisen, I. Damgaard, M. Chellakooty, I. Schmidt, H. Virtanen, D. Petersen, A. Andersson, J. Toppari and N. Skakkebaek, "Human Breast Milk Contamination with Phthalates and Alterations of Endogenous Reproductive Hormones in Infants Three Months of Age," *Environmental Health Perspectives* 114 (2006): 270–76.

38 R. Hauser, J. Meeker, S. Duty, M. Silva and A. Calafat, "Altered Semen Quality in Relation to Urinary Concentrations of Phthalate Monoester and Oxidative Metabolites," *Epidemiology* 17 (2006): 682–91; R. Hauser, J. Meerker, N. Singh, M. Silva, L. Ryan and S. Duty, "DNA Damage in Human Sperm Is Related to Urinary Levels of Phthalate Monoester and Oxidative Metabolites," *Human Reproduction* 22 (2007): 688–95.

39 Swan et al. "Prenatal Phthalate Exposure."

Chapter 2

1 Complete survey results were the following. Of the 245 surveys completed:

149 – 60.8% – (A) I am concerned about exposure to toxins in non-organic food.

17 – 6.9% – (B) I believe organic foods taste better.

31 – 12.7% – (C) I think that organic foods have more nutrients than non-organic foods.

19 – 7.8% – (D) I like to support local farmers.

29 – 11.8% – (E) I believe organic food is better for the environment.

2 S. Lockie, K. Lyons, G. Lawrence and J. Grice, "Choosing Organics: A Path Analysis of Factors Underlying the Selection of Organic Food among Australian Consumers,"*Appetite* 43 (2004): 135–46. Maryellen

Molyneaux, President, Natural Marketing Institute, "Engaging the Next Wave: An Organic Consumer Study" (presentation at Organic Trade Association—State of the Organic Industry, Baltimore, Maryland, September 2012).

3 C. Benbrook, "Initial Reflections on the *Annals of Internal Medicine* Paper 'Are Organic Foods Safer and Healthier Than Conventional Alternatives? A Systematic Review,'" 2012, www.tfrec.wsu.edu/pdfs/P2566.pdf.

4 C. Benbrook, "Simplifying the Pesticide Risk Equation: The Organic Option," The Organic Center, Boulder, CO, 2008, www.organiccenter.org/science.pest.php?action=view&report_id=125; C. Benbrook, "The Organic Center's 'Dietary Risk Index'—Tracking Relative Pesticide Risks in Food and Beverages," The Organic Center, Boulder, CO, 2011, www.organiccenter.org/reportfiles/DRIfinal_09--D10--D2011.pdf.

5 C. Lu, K. Toepel, R. Irish, R. Fenske, D. Barr and R. Bravo, "Organic Diets Significantly Lower Children's Dietary Exposure to Organophosphorous Pesticides," *Environmental Health Perspectives* 114 (2006): 260–63.

6 Rachel Carson, *Silent Spring* (New York: Houghton Mifflin, 1962).

7 C. Lu, D. Barr, M. Pearson, L. Walker and R. Bravo, "The Attribution of Urban and Suburban Children's Exposure to Synthetic Pyrethroid Insecticides: A Longitudinal Assessment," *Journal of Exposure Science and Environmental Epidemiology* 19 (2009): 69–78.

8 C. Lu, K. Warchol and R. Callahan, "*In situ* Replication of Honey Bee Colony Collapse Disorder," *Bulletin of Insectology* 65 (2012): 99–106.

9 P. Jeschke, R. Nauen, M. Schindler and A. Elbert, "Overview of the Status and Global Strategy for Neonicotinoids," *Journal of Agriculture and Food Chemistry* 59 (2011): 2897–2908.

10 R. Morse and N. Calderone, "The Value of Honey Bees as Pollinators of U.S. Crops in 2000," 2000, accessed November 21, 2012, http://www.masterbeekeeper.org/pdf/pollination.pdf; N. Gallai, J. Salles, J. Settele and B. Vaissière, "Economic Valuation of

the Vulnerability of World Agriculture Confronted with Pollinator Decline," *Ecological Economics* 68(2009): 810–21.

11 EMF Safety Network, "Cell Phone Radiation Disturbs Honey Bees," May 10, 2011, accessed November 21, 2012, http://emfsafetynetwork.org/?p=4028.

12 European Commission: Animal Health and Welfare, "Bees and Pesticides: Commission Goes Ahead with Plan to Better Protect Bees," May 1, 2013, accessed July 11, 2013, http://ec.europa.eu/food/animal/liveanimals/bees/neonicotinoids_en.htm.

13 United States Department of Agriculture: National Honey Bee Health Stakeholder Conference Steering Committee, *Report on the National Stakeholders Conference on Honey Bee Health*, 2012.

14 C. Smith-Spangler, M. Brandea, G. Hunter, J. Bavinger, M. Pearson, P. Eschbach, V. Sundaram, H. Liu, P. Schirmer, C. Stave, I. Olkin and D. Bravata, "Are Organic Foods Safer or Healthier Than Conventional Alternatives? A Systematic Review," *Annals of Internal Medicine* 157 (2012): 348–66.

15 Benbrook, "Initial Reflections on the *Annals of Internal Medicine* Paper."

16 K. Brandt, C. Leifert, R. Sanderson and C. Seal, "Agroecosystem Management and Nutritional Quality of Plant Foods: The Case of Organic Fruits and Vegetables," *Critical Reviews in Plant Sciences* 30 (2011): 1–2, 177–97.

17 As we were in this book's editing stages, Matt Holmes sent me some highlights from his organization's study of the Canadian organic marketplace. The stats indicate that the Stanford study couldn't have been as much of a disaster as the media predicted. Canada's organic market continued to grow in 2012, totalling C$3.7 billion for the year—a tripling in sales from just six years before. And ignoring the naysayers, 58 percent of all Canadians purchase organic products every week.

18 See U.S. Department of Agriculture, "The Organic Integrity Quarterly: National Organic Program Newsletter," October 2012,

http://www.ams.usda.gov/AMSv1.0/getfile?dDocName=STELP RDC5100909; Food Marketing Institute, "Natural and Organic Food Overview," 2008, accessed November 23, 2012, http://www.fmi.org/docs/media-backgrounder/natural_organic_foods.pdf?sfvrsn=2, 2008.

19 S. Strom, "Has 'Organic' Been Oversized?" *New York Times*, July 7, 2012, accessed November 3, 2012, http://www.nytimes.com/2012/07/08/business/organic-food-purists-worry-about-big-companies-influence.html?pagewanted=all&_r=0.

20 The Cornucopia Institute, *White Paper: The Organic Watergate*, 2012, accessed November 23, 2012, http://www.cornucopia.org/2012/05/the-organic-watergate-advocates-condemn-corruption-and-us-das-cozy-relationship-with-corporate-agribusinesses-in-organics-2/.

21 Stonyfield Farm, "Milestones," November 27, 2012, http://www.stonyfield.com/about-us/our-story-nutshell/milestones.

22 M. Hirschberg, Stonyfield Farm, "The Full Story," accessed November 27, 2012, http://www.stonyfield.com/about-us/our-story-nutshell/full-story.

23 K. McCormack, "Stonyfield CEO Resigns to Focus on Food Policy," *Bloomberg Businessweek*, January 12, 2012, accessed November 27, 2012, http://www.businessweek.com/ap/financialnews/D9S7KLKO1.htm.

24 M. Gunther, "Stonyfield Stirs Up the Yogurt Market," CNN Money Online, January 4, 2008, accessed November 27, 2012, http://money.cnn.com/2008/01/03/news/companies/gunther_yogurt.fortune/index.htm.

25 CNBC.com, "Global Food-Giant Buyouts of Top Organic Brands," October 8, 2012, accessed November 28, 2012, http://www.cnbc.com/id/49185619/Global_Food_Giant_Buyouts_of_Top_Organic_Brands.

26 U.S. Department of Agriculture, Economic Research Service, "2011 Foodstore Sales and Sales Growth, Share of Sales by Retail Segment and Industry Structure," 2012, accessed November 28, 2012, http://www.ers.usda.gov/topics/food-markets-prices/retailing-wholesaling/retail-trends.aspx.

27 A. Little, "An Interview with Walmart CEO H. Lee Scott,"*Grist*, April 13, 2006, accessed November 28, 2012, http://grist.org/business-technology/griscom-little3/.

28 V. Seufert, N. Ramankutty and J.Foley, "Comparing the Yields of Organic and Conventional Agriculture," *Nature* 485 (2012): 229–32.

29 California Right to Know, "Yes on Prop 37: Facts," accessed November 28, 2012, http://www.carighttoknow.org/facts.

30 Cornucopia Institute, "Consumer Guide to Pro/Con Prop 37 Brands," August 2012, accessed December 3, 2012, http://www.cornucopia.org/2012/08/agribusinesses-owning-naturalorganic-brands-betray-customers-fund-attack-on-gmo-labeling-proposal-in-california/.

31 Genetically modified organisms (GMOs) are organisms whose genetic material has been altered using genetic engineering (GE) techniques.

32 C. Benbrook, "Impact of Genetically Engineered Crop on Pesticide Use in the United States: The First Thirteen Years," The Organic Center, 2009, accessed November 28, 2012, https://www.organic-center.org/reportfiles/13Years20091126_ExSumFrontMatter.pdf.

33 California Right to Know, "Yes on Prop 37: Facts."

34 S. Finz, "Prop 37: Genetic Food Label Defeated," *San Francisco Chronicle*, November 7, 2012, accessed November 28, 2012, http://www.sfgate.com/news/article/Prop-37-Genetic-food-labels-defeated-4014669.php.

35 I know of only one other study in the scientific literature in addition to Lu's that has looked at this question. Cynthia Curl and her colleagues found that mean levels of certain organophosphate pesticides were nine times higher in preschool kids in Seattle who reported a conventional diet versus an organic one (C. Curl, R. Fenske and K. Elgethun, "Organophosphorus Pesticide Exposure of Urban and Suburban Preschool Children with Organic and Conventional Diets," *Environmental Health Perspectives* 111 [2003]: 377–82).

36 Curl et al., "Organophosphorus Pesticide Exposure"; Lu et al., "Organic Diets Significantly Lower Children's Dietary Exposure."

37 For our data analysis, we used the method as outlined in Curl et al., "Organophosphorus Pesticide Exposure"): All urine samples containing urine concentrations below the limit of detection (LOD) were assumed to have concentrations equal to one-half the LOD. Total dimethyl dialkylphosphate (DAP) molar quantities were calculated according to this formula:

$$[\text{Dimethyl DAP}] = [\text{DMP}]/125 + [\text{DMTP}]/141 + [\text{DMDTP}]/157$$

The data table we used showed the urinary dimethyl DAP metabolite concentrations in each of the nine participants on each day (Days 1 to 12). We then found an average of the nine participants' levels for each day and averaged those days into the three phases of the study. This left us with the three data points that you see on the graph.

38 L. Tan, N. Nielsen, D. Young and Z. Trizna for the Council on Scientific Affairs, "Use of Antimicrobial Agents in Consumer Products," *Archives of Dermatology* 138 (2002): 1082–86, accessed December 9, 2012, http://www.ama-assn.org/ama1/pub/upload/mm/443/csaa-00.pdf; Canadian Medical Association, "Antimicrobial/Antibacterial Products: Public Health Issue Briefing," 2010, accessed December 9, 2012, http://www.cma.ca/multimedia/CMA/Content_Images/Inside_cma/Office_Public_Health/HealthPromotion/Antimicrobial-IssueBriefing_en.pdf.

39 M. Sanborn, K. Bassil, C. Vakil, K. Kerr and K. Ragan, "2012 Systematic Review of Pesticide Health Effects," Ontario College of Family Physicians; Canadian Cancer Society, "Harmful Substances and Environmental Risks: Our Position on Cosmetic Use of Pesticides," accessed December 9, 2012, http://www.cancer.ca/en/prevention-and-screening/be-aware/harmful-substances-and-environmental-risks/pesticides/our-position/?region=on.

40 S. Reuben for the President's Cancer Panel, *2008–2009 Annual Report: Reducing Environmental Cancer Risk*, U.S. Department of Health and Human Services, National Institute of Health and the National

Cancer Institute (Bethesda, MD, 2010), accessed December 9, 2012, http://deainfo.nci.nih.gov/advisory/pcp/annualreports/pcp08-09rpt/PCP_Report_08-09_508.pdf.

41 The American Academy of Pediatrics released this statement (accessed November 26, 2012)—http://www.aap.org/en-us/about-the-aap/aap-press-room/Pages/American-Academy-of-Pediatrics-Weighs-In-For-the-First-Time-on-Organic-Foods-for-Children.aspx—in reference to the following study: J. Forman and J. Silverstein, "Organic Foods: Health and Environmental Advantages and Disadvantages," *Pediatrics* (2012), doi: 10.1542/peds.2012-2579.

42 E. Bräuner, M. Sorensen, E. Gaudreau, A. LeBlanc, K. Eriksen, A. Tionneland, K. Overvad and O. Raaschou-Nielsen, "Prospective Study of Organochlorines in Adipose Tissue and Risk of Non-Hodgkin's Lymphoma," *Environmental Health Perspectives* 121 (2012): 105–11.

43 F. Orton, E. Rosivatz, M. Scholze and A. Kotenkamp, "Widely Used Pesticides with Previously Unknown Endocrine Activity Revealed as In Vitro Androgens," *Environmental Health Perspectives* 119 (2011): 794–800.

44 M. Bouchard, D. Bellinger, R. Wright and M. Weisskopf, "Attention-Deficit/Hyperactivity Disorder and Urinary Metabolites of Organophosphate Pesticides," *Pediatrics* 125 (2010), no. 6; M. Bouchard, J. Chevier, K. Harley, K. Kogut, M. Vedar, N. Calderon, C. Trujillo, C. Johnson, A. Bradman, D. Barr and B. Eskenazi, "Prenatal Exposure to Organophosphate Pesticides and IQ in Seven-Year-Old-Children," E*nvironmental Health Perspectives* 19 (2011): 1189–95; S. Engel, J. Wetmur, J. Chen, C. Zhu, D. Barr, R. Canfield and M. Wolff, "Prenatal Exposure to Organophosphates, Paraoxonase 1 and Cognitive Development in Childhood," *Environmental Health Perspectives* 199 (2011): 1182–88; A. Marks, K. Harley, A. Broadman, K. Kogu, D. Barr, C. Johnson, N. Calderon and B. Eskenazi, "Organophosphate Pesticide Exposure and Attention in Young Mexican-American Children: The CHAMACOS Study," *Environmental Health Perspectives* 118 (2010):1768–74.

45 C. Chevrier, G. Limon, C. Monfort, F. Rouget, R. Garlantezec, C. Petit, G. Durand and S. Cordier, "Urinary Biomarkers of Prenatal

Atrazine Exposure and Adverse Birth Outcomes in the PELAGIE Birth Cohort," *Environmental Health Perspectives* 119 (2011): 1034–41.

46 S. Christiansen, M. Schloze, M. Dalgaard, A. Vinggaard, M. Axelstad, A. Koretenkamp and U. Hass, "Synergistic Disruption of External Male Sex Organ Development by a Mixture of Four Antiandrogens," *Environmental Health Perspectives* 117 (2009): 1839–46.

47 A. Hernandez, T. Parron and R. Alarcon, "Pesticides and Asthma," *Current Opinion in Allergy and Clinical Immunology* 11 (2011): 90–96.

48 A. Adigun, N. Wrench, F. Seidler and T. Slotkin, "Neonatal Organophosphate Pesticide Exposure Alters the Developmental Trajectory of Cell-Signalling Cascades Controlling Metabolism: Differential Effects of Diazinon and Parathion," *Environmental Health Perspectives* 118 (2010): 210–15; D. Lee, M. Steffes, A. Sjodin, R. Jones, L. Needham, D. Jacobs, "Low Dose Organochlorine Pesticides and Polychlorinated Biphenyls Predict Obesity, Dyslipedemia, and Insulin Resistance among People Free of Diabetes," *PLoS ONE* 6 (2011), no. 1.

49 N. Tadevosyan, "Pesticides Application in Agriculture of Armenia and Their Impact on Reproductive Function in Humans," *New Armenian Medical Journal 3 (2009): 41–48.*

Chapter 3

1 Amy Winehouse, "Rehab" (remix produced by Jay-Z), http://www.youtube.com/watch?v=cbF18itWXvc.

2 B. Dixon, "'Detox': A Mass Delusion," *The Lancet* 5 (2005): 261.

3 A reference to the 1976 film starring John Travolta, based loosely on the life of David Vetter, who was born with Severe Combined Immunodeficiency (SCI), a condition so dangerous that he literally lived inside a plastic tent to prevent any contact with germs or viruses in the outside world. Sadly, even going to that extent, it was not possible to keep him totally isolated from germs, and he died in 1984.

4 Marketdata Enterprises, a U.S.-based independent market research firm, released a report in 2011 looking at the weight loss and dieting

market in the United States (Marketdata Enterprises Inc., *U.S. Weight Loss and Diet Control Market*, 11th ed. (Tampa, FL: Marketdata Enterprises Inc., 2011). According to their research, the sales of diet and detox pills alone total over US$2.5 billion a year in the United States. Another market intelligence firm, Euromonitor International, looked at the sales of top antioxidant ingredient sales (things like green tea, super-fruit juice and dietary supplements) and used those figures to estimate the combined global sales in this category: a total of US$34 billion in 2010 (Euromonitor International, *Health and Wellness: Global Briefing Series* [London, UK: Euromonitor International, 2011]). And those massive figures don't factor in many of the products and techniques we're about to examine.

5 For a detailed description of the science of chelation see B. Halstead, *The Scientific Basis of EDTA Chelation Therapy* (Colton, CA: Golden Quill Publishers, 1979).

6 E. Ernst, "Colonic Irrigation and the Theory of Autointoxication: A Triumph of Ignorance over Science," *Journal of Clinical Gastroenterology* 24 (1997), no. 4: 196–98.

7 MobiThinking, "Mobile Marketing Experts: Global Mobile Statistics," 2013, accessed March 2013, http://mobithinking.com/mobile-marketing-tools/latest-mobile-stats/a#subscribers.

8 S. Genuis, "Fielding a Current Idea: Exploring the Public Health Impact of Electromagnetic Radiation," *Public Health* 122 (2008): 113–24. See also the recent campaign ranking the radiation output of various cellphones undertaken by our friends at the Environmental Working Group (www.ewg.org).

9 For more information, check out the Happiness Project online: http://www.projecthappiness.org/programs/the-science-of-happiness/.

10 For an excellent account of the contamination of human breasts, see F. Williams, *Breasts: A Natural and Unnatural History* (New York: W.W. Norton, 2012).

11 U.S. Department of Veterans Affairs, "Public Health—Military Exposures at Camp Lejeune," accessed April 2013, http://www.publichealth.va.gov/exposures/camp-lejeune/.

12 "Camp Lejeune Lawsuit Goes Forward," *Veterans Today Military & Foreign Affairs Journal*, February 4, 2011, accessed June 2013, http://www.veteranstoday.com/2011/02/04/camp-lejeune-lawsuit-goes-forward/.

13 For the full story, watch the 2011 documentary film *Semper Fi: Always Faithful*.

14 U.S. National Cancer Institute, "Cancer Stats Fact Sheet," accessed April 2013, http://seer.cancer.gov/statfacts/.

15 American Cancer Society, "Cancer Facts and Figures 2010," Atlanta, GA, 2010, accessed April 2013, http://www.cancer.org/acs/groups/content/@nho/documents/document/acspc-024113.pdf.

16 Canadian Cancer Society, Canadian Cancer Society's Steering Committee on Cancer Statistics, Toronto, ON, Canadian Cancer Statistics 2012, accessed April 2013, http://www.cancer.ca/~/media/cancer.ca/CW/cancer%20information/cancer%20101/Canadian%20cancer%20statistics/Canadian-Cancer-Statistics-2012---English.pdf.

17 Australian Institute of Health and Welfare, Canberra, "Cancer in Australia 2010: An Overview," accessed April 2013, http://www.aihw.gov.au/publication-detail/?id=6442472459.

18 Kathy Freston, "A Vegan Diet (Hugely) Helpful against Cancer," Huffington Post Canada, December 9, 2012, accessed April 2013, http://www.huffingtonpost.com/kathy-freston/vegan-diet-cancer_b_2250052.html.

19 M. Pollan, *Food Rules: An Eater's Manual* (New York: Penguin Books, 2009).

20 Excellent resources exist for anyone wishing to delve into the details of diets, cleanses and detox. Some are written by people telling their personal detox stories; others are written by doctors, nurses, homeopaths and naturopaths. *The Non-Toxic Avenger* (a personal detox story

inspired by *Slow Death by Rubber Duck*), *Detoxify or Die*, *Detox for the Rest of Us*, *The Detox Strategy* and *Wylde on Health* are some of the many available. Check out some of these: D. Duke, *The Non-Toxic Avenger: What You Don't Know Can Hurt You* (Gabriola Island, BC: New Society, 2011); S.A. Rogers, *Detoxify or Die* (Sarasota, FL: Sand Key Company, 2002); C. Jacobs, *Detox for the Rest of Us* (Avon, MA: Adams Media, 2010); B. Watson, *The Detox Strategy* (New York: Free Press, 2008); B. Wylde, *Wylde on Health: Your Best Choices in the World of Natural Health* (Toronto: Random House, 2012).

21 See Centers for Disease Control and Prevention, "Colorectal Cancer Statistics," accessed April 2013, http://www.cdc.gov/cancer/colorectal/statistics/, and Canadian Cancer Society, "Colorectal Cancer Statistics," accessed April 2013, http://www.cancer.ca/en/cancer-information/cancer-type/colorectal/statistics/?region=on.

22 T.H. Risby and S.F. Solga, "Current Status of Clinical Breath Analysis," *Applied Physics B* 85 (2006): 421–26.

23 Ibid.

24 Presence of birth control drugs in exhaled breath mentioned by Dr. Stephen Genuis in interview with author in August 2012.

25 Harold and Arline Brecher, *Forty Something Forever: A Consumer's Guide to Chelation Therapy and Other Heart Savers* (Herndon, VA: Heart Savers Press, 1997).

26 C. Lamas, C. Goertz, R. Boineau, D. Mark, T. Rozema, R. Nahin, L. Lindlblad, E. Lewis, J. Drisko and K. Lee, "Effect of Disodium EDTA Chelation Regimen on Cardiovascular Events in Patients with Previous Myocardial Infarction: The TACT Randomized Trial," *The Journal of the American Medical Association* 309 (2013): 1241–250.

27 R. Jandacek, N. Anderson, M. Liu, S. Zheng, Q. Yang and P. Tso, "Effects of Yo-Yo Diet, Caloric Restriction and Olestra on Tissue Distribution of Hexachlorobenzene," *American Journal of Physiology and Gastrointestinal Liver Physiology* 288 (2005): G292-G299; M. Imamura and T. Tung, "A Trial of Fasting Cure for PCB-Poisoned Patients in Taiwan," *American Journal of Industrial Medicine* 5 (1985): 147–53.

28 M. Blanusa, V. Varnai, M. Piasek and K. Kostial, "Chelators as Antidotes of Metal Toxicity: Therapeutic and Experimental Aspects," *Current Medical Chemistry* 12 (2005): 2771–94; H. Aposhian, R. Maiorino, D. Gonzalez-Ramirez, M. Zuniga-Charles, Z. Xu, K. Hurlbut, P. Junco-Munoz, R. Dart and M. Aposhian, "Mobilization of Heavy Metals by Newer, Therapeutically Useful Chelating Agents," *Toxicology* 97 (1995): 23–38.

29 D. Kennedy, K. Cooley, T. Einarson and D. Seely, "Objective Assessment of an Ionic Footbath (IonCleanse): Testing Its Ability to Remove Potentially Toxic Elements from the Body," *Journal of Environmental and Public Health* 2012 (2012): 1–13.

30 S. Horne, "Colon Cleansing: A Popular, but Misunderstood Natural Therapy," *Journal of Herbal Pharmacotherapy* 6 (2006): 93–100; R. Acosta and B. Cash, "Clinical Effects of Colonic Cleansing for General Health Promotion: A Systematic Review," *American Journal of Gastroenterology* 104 (2009): 2830–36.

31 V. Nadeau, G. Truchon, M. Brochu and R. Tardif, "Effect of Physical Exertion on the Biological Monitoring of Exposure to Various Solvents following Exposure by Inhalation in Human Volunteers: I. Toluene," *Journal of Occupational Environmental Hygiene* 3 (2006): 481–89; R. Tardif, V. Nadeau, G. Truchon and M. Brochu, "Effect of Physical Exertion on the Biological Monitoring of Exposure to Various Solvents following Exposure by Inhalation in Human Volunteers: II. N-Hexane," *Journal of Occupational Environmental Hygiene* 4 (2007): 502–508; quiz D568-569.

32 F. Ibrahim, T. Halttunen, R. Tahvonen and S. Salminen, "Probiotic Bacteria as Potential Detoxification Tools: Assessing Their Heavy Metal Binding Isotherms," *Canadian Journal of Microbiology* 52 (2006): 877–85.

33 A. Barnes, M. Smith, S. Kacinko, E. Schwilke, E. Cone, E. Moolchan and M. Huestis, "Excretion of Methamphetamine and Amphetamine in Human Sweat following Controlled Oral Methamphetamine Administration," *Clinical Chemistry* 54 (2008): 172–80; N. Fucci, N. de Giovanni and S. Scarlata, "Sweat Testing in Addicts under Methadone

Treatment: An Italian Experience," *Forensic Science International* 174 (2008): 107–10.

34 J. Domingo, M. Gomez, J. Llobet and J. Corbella, "Comparative Effects of Several Chelating Agents on the Toxicity, Distribution and Excretion of Aluminum," *Human Toxicology* 7 (1998): 259–62; Z. Zhao, L. Liang, X. Fan et al. "The Role of Modified Citrus Pectin as an Effective Chelator of Lead in Children Hospitalized with Toxic Lead Levels," *Alternative Therapeutic Health Medicine* 14 (2008): 34–38.

35 Adapted from S. Genuis, "Elimination of Persistent Toxicants from the Human Body," *Human and Experimental Toxicology* 30 (2011): 3–18, with added language to simplify the detox effectiveness column.

Chapter 4

1 M. Squibb, *Lipid Detoxification: Clearing Fat-Soluble Toxins from Cellular Lipid Structures,* Whole Health Research Alliance, 2008, accessed April 2013, http://ebooks.whnlive.com/LipophilicDetox/LipidDetox.pdf.

2 Choice Online, October 2005, accessed April 2013, http://www.choice.com.au/reviews-and-tests/food-and-health/general-health/therapies/detox-kits-review-and-compare/page.aspx.

3 Gerard E. Mullin, "Popular Diets Prescribed by Alternative Practitioners—Part 1," *Nutrition in Clinical Practice* 25 (2010), no. 2: 212–14.

4 D. Valle and T. Manolio, "Applying Genomics to Clinical Problems—Diagnostics, Preventive Medicine, Pharmacogenomics: A White Paper for the National Human Genome Research Institute," National Institutes of Health, 2008, accessed June 27, 2013, http://www.genome.gov/27529204.

5 Stanford Medicine Cancer Institute, "Hereditary Breast Ovarian Cancer Syndrome (BRCA1/BRCA2)," 2013, accessed June 4, 2013, http://cancer.stanford.edu/information/geneticsAndCancer/types/herbocs.html.

6 A. Jolie, "My Medical Choice," *New York Times*, May 14, 2013, accessed June 4, 2013, http://www.nytimes.com/2013/05/14/opinion/my-medical-choice.html.

7 U.S. Breast Cancer Statistics, BreastCancer.Org, 2012, accessed June 4, 2013, http://www.breastcancer.org/symptoms/understand_bc/statistics.

8 T. Doshi, S.S. Mehta, V. Dighe, N. Balasinor and G. Vanage, "Hypermethylation of Estrogen Receptor Promoter Region in Adult Testis of Rats Exposed Neonatally to Bisphenol A," *Toxicology* (2011), http://dx.doi.org/10.1016/j.tox.2011.07.011.

9 H. Roberts, "Vitamin C: Linus Pauling Was Right All Along—A Doctor's Opinion," August 17, 2004, accessed April 2013, http://www.medicalnewstoday.com/releases/12154.php.

10 *The Dr. Oz Show*, "5 Foods to Prevent Your Arteries from Clogging," accessed August 2013, http://www.doctoroz.com/videos/5-foods-prevent-your-arteries-clogging.

11 R.J. Reiter, D.X. Tan, L.C. Manchester, S. Lopez-Burillo, R.M. Sainz and J.C. Mayo, "Melatonin: Detoxification of Oxygen and Nitrogen-Based Toxic Reactants," *Advances in Experimental Medicine and Biology* 527 (2003): 539–48.

12 S.J. James, W. Slikker III, S. Melnyk, E. New, M. Pogribna and S. Jernigan, "Thimerosal Neurotoxicity Is Associated with Glutathione Depletion: Protection with Glutathione Precursors," *NeuroToxicology* 26 (2005), no. 1: 1–8.

13 L. Lands, V. Grey and A. Smountas, "Effect of Supplementation with a Cysteine Donor on Muscular Performance," *Journal of Applied Physiology* 87 (1999): 1381–85.

14 Ibid.

15 National Fire Protection Association, "Statistics: Fires by Type," accessed March 2013, http://www.nfpa.org/itemDetail.asp?categoryID=953&itemID=52970&URL=Research/Fire%20statistics/The%20U.S.%20fire%20problem.

16 C.C. Austin, D. Wang, D.J. Ecobichon and G. Dussault, "Characterization of Volatile Organic Compounds in Smoke at Environmental Fires," *Journal of Toxicology and Environmental Health* 63 (2001): 437–58.

17 M. Webber, J. Gustave, R. Lee, J.K. Niles, K. Kelly, H.W. Cohen and D. Prezant, "Trends in Respiratory Symptoms of Firefighters Exposed to the World Trade Center Disaster: 2001–2005," *Environmental Health Perspectives* 117 (2009): 975–80.

18 J. Li, J. Cone, A. Kahn, R. Brackbill, M. Farfel, C. Greene, J. Hadler, L. Stayner and S. Stellman, "Association between World Trade Center Exposure and Excess Cancer Risk," *Journal of the American Medical Association* 308 (2012): 2479–88.

19 New Hampshire Department of Environmental Services, "Environmental Health Program: Cancer and the Environment," accessed March 2013, http://des.nh.gov/organization/commissioner/pip/publications/co/documents/cancer_environment.pdf.

20 J. Shatki, "Wellness Education: Ayurveda," 2010, accessed March 2013, https://www.jivanshakti.com/ayurveda.aspx.

21 J. Dahl and K. Falk, "Ayurvedic Herbal Supplements as an Antidote to 9/11 Toxicity," *Alternative Therapies in Health and Medicine* 14 (2008): 24–28.

22 My intention is not to minimize the value of the study but to point out that the data is survey data, not clinical testing data of patients' chemical body burdens. A study of declines in their chemical levels would be very helpful, however.

23 Dahl and Falk, "Ayurvedic Herbal Supplements as an Antidote to 9/11 Toxicity."

24 S. Genuis, "Blood, Urine and Sweat (BUS) Study: Monitoring and Elimination of Bioaccumulated Toxic Elements," *Archives of Environmental Contamination and Toxicology* 61 (2011): 344–57.

25 We wish to thank AXYS Labs of Sidney, BC, for their advice and support in undertaking the sweat analysis.

26 S. Genuis, "Elimination of Persistent Toxicants from the Human Body," *Human and Experimental Toxicology* 30 (2010): 3–18; Genuis, "Blood, Urine and Sweat (BUS) Study: Monitoring and Elimination of Bioaccumulated Toxic Elements."

27 According to the claims of some infrared sauna manufacturers, it is possible to produce a litre of sweat in 15 minutes. This is highly unlikely, but based on my experience, it could be possible to produce this much sweat during a 45-minute session.

28 See Table 1 in A. Calafat, Z. Kuklenvik, J. Reidy, S. Caudill, J. Ekong and L. Needham, "Urinary Concentrations of Bisphenol A and 4-Nonylphenol in a Human Reference Population," *Environmental Health Perspectives* 113 (2004): 391–95.

29 See Table 1 in R. Stahlhut, E. van Wijngaarden, T. Dye, S. Cook and S. Swan, "Concentrations of Urinary Phthalate Metabolites Are Associated with Increased Waist Circumference and Insulin Resistance in Adult U.S. Males," *Environmental Health Perspectives* 115 (2007): 876–82.

30 Genuis, "Blood, Urine and Sweat (BUS) Study: Monitoring and Elimination of Bioaccumulated Toxic Elements."

Chapter 5

1 AOL Autos, "Chevrolet Tahoe: Model Overview," 2013, accessed February 11, 2013, http://autos.aol.com/cars-Chevrolet-Tahoe-2013/overview/.

2 For example, the Little Trees Car Freshner Corporation New Car Scent, accessed February 2013, http://www.car-freshner.com/little-trees-air-fresheners/new-car-scent-little-trees-air-freshener.

3 J. Ramsey, "What's That Smell? It's a Bentley Boy,"AOL Autoblog, February 24, 2013, accessed February 2013, http://www.autoblog.com/2013/02/24/whats-that-smell-its-a-bentley-boy/#continued.

4 C. Clover, "Enjoying the Smell of a New Car 'Is Like Sniffing Glue,'" *The Telegraph*, January 15, 2003, accessed February 2013, http://www.

telegraph.co.uk/news/worldnews/asia/japan/1418964/Enjoying-the-smell-of-a-new-car-is-like-glue-sniffing.html.

5 Ecology Center, *2011/2012 Guide to New Vehicles*, 2012, accessed February 13, 2013, http://www.healthystuff.org/documents/2012_Cars.pdf.

6 N. Klepeis, W. Nelson, J. Robinson, A. Tsang, P. Switzer, J. Behar, S. Hern and W. Engelman, "The National Human Activity Pattern Survey (NHAPS): A Resource for Assessing Exposure to Environmental Pollutants," *Journal of Exposure Analysis and Environmental Epidemiology* 11 (2001): 231–52. See also U.S. EPA Indoor Air Division and Office of Research and Development, *Volume 1: Federal Programs Addressing Indoor Air Quality*, 1989, http://1.usa.gov/XfwS7G.

7 C.R. Kirman, L.L. Aylward, B.C. Blount, D.W. Pyatt and S.M. Hays, "Evaluation of NHANES Biomonitoring Data for Volatile Organic Compounds in Blood: Application of Chemical-Specific Screening Criteria," *Journal of Exposure Science and Epidemiology* 22 (2012): 24–34.

8 This study found that levels of some brominated pollutants are up to 10 times the maximum level found in homes and offices: C.C. Carignan, M. D.McClean, E.M. Cooper, D.J. Watkins, A.J. Fraser, W. Heiger-Bernays, H.M. Stapleton and T.F. Webster, "Predictors of Tris(1,3-dichloro-2-propyl) Phosphate Metabolite in the Urine of Office Workers," *Environment International* 55 (2013) 56–61. And these studies found a myriad of chemicals in the cabin air of automobiles: T. Yoshida, "Approach to Estimation of Absorption of Aliphatic Hydrocarbons Diffusing from Interior Materials in an Automobile Cabin by Inhalation Toxicokinetic Analysis in Rats," *Journal of Applied Toxicology* 30 (2009): 42–52; O. Geiss, S. Tirendi, J. Barrero-Moreno, D. Kotzias, "Investigation of Volatile Organic Compounds and Phthalates Present in the Cabin Air of Used Private Cars," *Environment International* 35 (2009): 1188–95.

9 L. Sabatini, A. Barbieri, P. Indiveri, S. Mattioli and F.S. Violante, " Validations of an HPLC-MS/MS Method for the Simultaneous Determination of Phenylmercapturic Acid, Benzylmercapturic Acid and *O*-methylbenzyl Mercapturic Acid in Urine as Biomarkers of Exposure to Benzene, Toluene and Xylenes," *Journal of Chromatography* 863 (2008):

115–22; J.G. Filser, G.A. Csanady, W. Dietz, W. Kessler, P.E. Krenzen, M. Richter and A. Stormer, "Comparative Estimation of the Neurotoxic Risks of N-Hexane and N-Heptane in Rats and Humans Based on Formation of Metabolites 2,5-Hexanedione and 2,5-Heptanedione,"*Advances in Experimental Medicine and Biology* 387 (1996): 411–27.

10 J.T. Brophy, M. Keith, A. Watterson, R. Park, M. Gilbertson, E. Maticka-Tyndale, M. Beck, H. Abu-Zahra, K. Schneider, A. Reinhartz, R. DeMatteo and I. Luginaah, "Breast Cancer Risk in Relation to Occupations with Exposure to Carcinogens and Endocrine Disruptors: A Canadian —Case-Control Study," *Environmental Health* 11 (2012). doi:10.1186/1476-069X-11-87.

11 R. DeMatteo, M. Keith, J.T. Brophy, A. Wordsworth, A. Watterson, M. Beck, A. Ford, M. Gilberston, J. Pharityal, M. Rootham and D. Scott, "Chemical Exposure of Women Workers in the Plastics Industry with Particular Reference to Breast Cancer and Reproductive Hazards," *New Solutions* 22 (2012): 427–48.

12 Ibid.

13 Wikipedia entry, "Injection Molding," accessed March 2013, http://en.wikipedia.org/wiki/Injection_molding#Applications.

14 Brophy et al., "Breast Cancer Risk."

15 Ibid.

16 Ibid. Also recall from Chapter 3 the high incidence of breast cancer in men who had lived at Camp Lejeune in North Carolina. The water at this military base had been contaminated with industrial chemicals like perchloroethylene (PCE) and trichloroethylene (TCE), which are used to degrease the type of heavy machinery used in plastics manufacturing.

17 W. Wang, Y. Qiu, J. Jiao, J. Liu, F. Ji, W. Miao, Y. Zhu and Z. Xia, "Genotoxicity in Vinyl-Chloride Exposed Workers and Its Implications for Occupational Exposure Limits," *American Journal of Industrial Medicine* 54 (2011): 800–810.

18 C. Martinez-Valenzuela, S. Gomez-Arroyo, R. Villalobos-Pietrini, S. Waliszewski, M. Calderón-Segura, R. Félix-Gastélum and A. Alvarez-

Torres, "Genotoxic Biomonitoring of Agricultural Workers Exposed to Pesticides in the North of Sinaloa State, Mexico," *Environment International* 35 (2009): 1155–59.

19 C. Regina dos Santos, M. Meye Passarelli and E. de Souza, "Evaluation of 2,5-Hexanedione in Urine in Workers Exposed to N-Hexane in Brazilian Shoe Factories," *Journal of Chromatography* 778 (2002): 237–44.

20 L. Costa and G. Giordano, "Developmental Neurotoxicity of Polybrominated Diphenyl Ether (PBDE) Flame Retardants," *Neurotoxicology* 28 (2007): 1047–67; C. Dufault, G. Poles and L. Driscoll, "Brief Postnatal PBDE Exposure Alters Learning and the Cholinergic Modulation of Attention in Rats," *Toxicological Sciences* 88 (2005): 172–80; K. Harley, J. Chevrier, R.A. Schall, A. Sjodin, A. Bradman and B. Eskenazi, "Association of Prenatal Exposure to PDEs and Infant Birth Weight," *American Journal of Epidemiology* 174 (2001): 1–8; E. Roze, L. Meijer, A. Bakker and K. Van Braeckel, "Prenatal Exposure to Organohalogens, including Brominated Flame Retardants, Influences Motor, Cognitive and Behavioral Performance at School Age," *Environmental Health Perspectives* 117 (2009): 1953–58.

21 L. Zhu and R.A Hites, "Temporal Trends and Spatial Distributions of Brominated Flame Retardants in Archived Fiches from the Great Lakes," *Environmental Science and Technology* 38 (2004): 2779–84.

22 A. Schecter, O. Papke, K. Tung, J. Joseph, T. Harris and J. Dahlgren, "Polybrominated Diphenyl Ether Flame Retardants in the U.S. Population: Current Levels, Temporal Trends and Comparison with Dioxins, Dibenzofurans and Polychlorinated Biphenyls," *Journal of Occupational and Environmental Medicine* 47 (2005): 199–211.

23 C. Butt, M. Diamond, J. Truong, M. Ikonomov and A. Ter Schure, "Spatial Distribution of PBDEs in Southern Ontario as Measured in Indoor and Outdoor Window Organic Films," *Environmental Science and Technology* 38 (2004): 724–31.

24 H. Jones-Otazo, J. Clarke, M. Diamond, J. Archbold, G. Ferguson, T. Harner, G. Richardson, J. Ryan and B. Wilford, "Is House Dust the

Missing Exposure Pathway of PBDEs? An Analysis of the Urban Fate and Human Exposure to PBDEs," *Environmental Science and Technology* 39 (2005): 5121–30.

25 M. Fang, T. Webster, D. Gooden, M. Coopers, M. McClean, C. Carignan, C. Makey and H. Stapleton, " Investigating a Novel Flame Retardant Known as v6: Measurements in Baby Products, House Dust and Car Dust," *Environmental Science and Technology* 47 (2013): 4449–54.

26 Ibid.

27 J. Herbstman, A. Sjodid, M. Kurzon, S. Lederman, R. Jones, V. Rauh, L. Needham, D. Tang, M. Niedzwiecki, R. Wang and F. Perera, "Prenatal Exposure to PBDEs and Neurodevelopment," *Environmental Health Perspectives* 118 (2010): 712-19.

28 C.G. Bornehag and E. Nanberg, "Phthalate Exposure and Asthma in Children," *International Journal of Andrology* 33 (2010): 333–45; C.-G. Bornehag, J. Sundell, C.J. Weschler, T. Sigsgaard, B. Lundgren, M. Hasselgren and L. Hägerhed-Engman, "The Association between Asthma and Allergic Symptoms in Children and Phthalates in House Dust: A Nested Case-Control Study" *Environmental Health Perspectives* 112 (2004): 1393–97; R.E. Dodson, M. Nishioka, L.J. Standley, L.J. Perovich, J.G. Brody and R.A. Rudel, "Endocrine Disruptors and Asthma-Associated Chemicals in Consumer Products," *Environmental Health Perspectives* 120 (2012):935–43.

29 E. Woods, U. Bhaumik, S. Sommer, S. Ziniel, A. Kessler, E. Chan, R. Wilkinson, M. Sesma, A. Burack, E. Klements, L. Queenin, D. Dickerson and S. Nethersole, "Community Asthma Initiative: Evaluation of a Quality Improvement Program for Comprehensive Asthma Care," *Pediatrics* 129 (2012): 464–572.

30 R. Knox, "To Control Asthma, Start with the Home instead of the Child," NPR Health Blogs, March 18, 2013, accessed March 2013, http://www.npr.org/blogs/health/2013/03/18/174393981/to-control-asthma-start-with-the-home-instead-of-the-child.

31 Ibid. And for more details, read the report in Woods et al., "Community Asthma Initiative."

32 C.J. Weschler, "Changes in Indoor Pollutants since the 1950s," *Atmospheric Environment* 43 (2009): 156–72.

33 C.J. Weschler and W.W. Nazaroff, "SVOC Exposure Indoors: Fresh Look at Dermal Pathways," *Indoor Air* 22 (2012): 356–77.

34 Ibid.

35 Ibid.

36 Weschler, "Changes in Indoor Pollutants."

37 Canada Green Building Council, LEED Program, accessed March 2013, http://www.cagbc.org/AM/Template.cfm?Section=LEED.

38 U.S. Green Building Council, "March 2013 Monthly LEED Overview and Update," 2013, accessed March 2013, http://new.usgbc.org/resources/list/presentations.

39 U.S. Green Building Council, "Square Footage of LEED-Certified Existing Buildings Surpasses New Construction," press release, Washington, D.C., December 7, 2011, accessed March 2013, http://www.usgbc.org/ShowFile.aspx?DocumentID=10712.

40 Living Building Challenge website, accessed February 2013, http://living-future.org/lbc.

41 Canada Green Building Council, Living Building Challenge, accessed February 2013, http://www.cagbc.org/Content/NavigationMenu/Programs/LivingBuildingChallenge/default.htm.

42 J. Hiskes, "Google Drops Red List Building Materials, Vendors Listen Up," *Sustainable Industries*, May 2, 2011, accessed February 2013, http://sustainableindustries.com/articles/2011/04/google-drops-red-list-building-materials-vendors-listen.

43 Canadian Broadcasting Corporation (CBC), "Burned: Company Statements," *Marketplace*, November 23, 2012, accessed March 2013, http://www.cbc.ca/marketplace/episodes/2012/11/company-statements-4.html.

Chapter 6

1 Letter to the Editor, "The London Pharmacopoeia," *The Water Cure Journal and Hygienic Magazine* (June 1850), no. 35.

2 S. Xu, H. Zhang, P. He and L. Shao, "Leaching Behaviour of Bisphenol A from Municipal Solid Waste into Landfill Environment," *Environmental Technology* 32 (2011): 1269–77; E. Davis, S. Klosterhaus and H. Stapleton, "Measurement of Flame Retardants and Triclosan in Municipal Sewage Sludge and Biosolids," *Environment International* 40 (2012): 1–7.

3 K. Xia, A. Bhandari, K. Das and G. Dillar, "Occurrence and Fate of Pharmaceutical and Biosolids," *Journal of Environmental Quality* 34 (2005): 91–104.

4 R. Yang, H. Wei, J. Guo and A. Li, "Emerging Brominated Flame Retardants in the Sediment of the Great Lakes," *Environmental Science and Technology* 46 (2012): 3119–26.

5 D. Kolpin, E. Furlong, M. Meyer, E. Thurman, S. Zaugg, L. Barber and H. Buxton, "Pharmaceuticals, Hormones and Other Organic Wastewater Contaminants in U.S. Lakes and Streams, 1999–2000: A National Reconnaissance," *Environmental Science and Technology* 36 (2002): 1202–11.

6 H. Spiegelman, "Unintended Consequences: A Short History of Waste" (presentation at the Coast Waste Management Association Spring Conference, Victoria, BC, March 29, 2007).

7 D. Hoornweg and P. Bhada-Tata, "What a Waste: A Global Review of Solid Waste Management," World Bank, Urban Development Series Knowledge Papers, 2012.

8 Ibid.

9 National Oceanic and Atmospheric Administration, "Marine Debris Program Factsheet," accessed March 2013, http://marinedebris.noaa.gov/info/pdf/plastic.pdf.

10 "High Level of Plastics Found in Great Lakes," *Toronto Star*, March 12, 2013.

11 Avery-Gomm, P. O'Hara, L. Kleine, V. Bowes, L. Wilson and K. Barry,

"Northern Fulmars as Biological Monitors of Trends of Plastic Pollution in the Eastern North Pacific," *Marine Pollution Bulletin* 64 (September 2012), no. 9: 1776–8.1

12 Tim Cook, CEO, Apple, "Keynote Presentation (presentation at the Goldman Sacks Technology and Internet Conference, San Francisco, CA, February 15, 2012).

13 *China Business News*, May 2012, accessed March 2013, www.china.org.cn.

14 U.S. Environmental Protection Agency, "Waste Education Resources: The Life Cycle of a Cell Phone," 2012, accessed March 2013, http://www.cpa.gov/osw/education/pdfs/life-cell.pdf.

15 IDC Market Research Firm, accessed March 2013, http://www.electronicstakeback.com/wp-content/uploads/Facts_and_Figures_on_EWaste_and_Recycling.pdf.

16 J. Greene, "The Environmental Pitfalls at the End of an iPhone's Life," *CNET*, September 26, 2012, accessed March 2013, http://news.cnet.com/8301-13579-3-57520123-37/the-environmental-pitfalls-at-the-end-of-an-iphones-life/.

17 "Lead Levels in Children Linked to Rise in E-Waste Profits," *China Daily*, November 16, 2011, accessed March 2013, http://www.chinadaily.com.cn/cndy/2011-11/16/content_14101761.htm.

18 The industry is represented by Call2Recycle Canada. The board of directors includes representation from Sony, Rayovac, Energizer Canada Inc., Panasonic Canada and Procter & Gamble Canada.

19 Minnesota Pollution Control Agency, *2005–2008 Perfluorochemical Evaluation at Solid Waste Facilities in Minnesota: Technical Evaluation and Regulatory Management Approach*, April 14, 2010.

20 K. Ward, Jr., "C8 Linked to Thyroid, Bowel Disease," *The Charleston Gazette*, July 30, 2012.

21 P. Anastas and J. Warner, *Green Chemistry: Theory and Practice* (New York: Oxford University Press, 1998).

22 Ibid.

23 J.M. Benyus, *Biomimicry: Innovation Inspired by Nature* (New York: William Morrow, 1997).

24 City of Toronto, "Water Quality Assurance Program: Protecting Water Quality," accessed February 2013, http://www.toronto.ca/water/protecting_quality/quality_assurance.htm.

25 W. MacKenzie, N. Hoxie, M. Proctor, M.S. Gradus, K. Blair, D. Peterson, J. Kazmierczak, D. Addiss, K. Fox, J. Rose and J. Davis, "A Massive Outbreak in Milwaukee of Cryptosporidium Infection Transmitted through the Public Water Supply," *New England Journal of Medicine* 331 (1994): 161–67.

26 Environmental Protection Agency, *Cryptosporidium: Drinking Water Health Advisory* EPA-822-R-01-009, Washington, D.C., March 2001.

27 Hon. D. O'Connor, Ontario Ministry of the Attorney General, *Walkerton Inquiry: The Events of May 2000 and Related Issues*, 2002, accessed February 22, 2013, http://www.attorneygeneral.jus.gov.on.ca/english/about/pubs/walkerton/.

28 CBC, "Consumer Background: Bottled Water," *CBC News*, August 20, 2008, accessed January 17, 2013, http://www.cbc.ca/news/background/consumers/bottled-water.html.

29 Canada, Health Canada, "Environmental and Workplace Health: Drinking Water Quality," accessed February 22, 2013, http://www.hc-sc.gc.ca/ewh-semt/water-eau/drink-potab/index-eng.php.

30 World Health Organization, "Pharmaceuticals in Drinking Water," 2012, accessed March 2013, http://apps.who.int/iris/bitstream/10665/44630/1/9789241502085_eng.pdf.

31 Ibid.

32 E. Teuten, J. Saquing, D. Knappe, M. Barlaz, S. Jonsson, A. Björn, S. Rowland et al., "Transport and Release of Chemicals from Plastics to the Environment and to Wildlife," *Philosophical Transactions of the Royal Society B* 364 (2009): 2027–45.

33 Canada, Health Canada, "Food and Nutrition: Frequently Asked Questions about Bottled Water," accessed February 2013, www.hc-sc.gc.ca/fn-an/securit/facts-faits/faqs_bottle_water-eau_embouteillee-eng.php.

34 Ibid.

35 A. Theen, "Ivy Colleges Shunning Bottled Water Jab at $22 Billion Industry," *Bloomberg News*, March 7, 2012, accessed February 2013, http://www.bloomberg.com/news/2012-03-07/ivy-colleges-shunning-bottled-water-jab-at-22-billion-industry.html.

36 S. Didier, "Water Bottle Pollution Facts," *National Geographic*, 2011, accessed March 2013, http://greenliving.nationalgeographic.com/water-bottle-pollution-2947.html.

37 National Resources Defense Council (NRDC), "Water Issues: Bottled Water," accessed February 2013, http://www.nrdc.org/water/drinking/qbw.asp.

38 R. Copes, G. Evans and S. Verhille, "Bottled vs. Tap Water," *BC Medical Journal* 51 (2009): 112–13, accessed February 2013, http://www.bcmj.org/council-health-promotion/bottled-vs-tap-water.

39 *Consumer Reports: Water Filter Buying Guide,* July 2012, accessed February 2013, http://www.consumerreports.org/cro/water-filters/buying-guide.htm?pn=0.

40 NSF International, "About NSF," accessed February 2013, http://www.nsf.org/business/about_NSF/.

41 NSF International, "Drinking Water Treatment Standards," accessed February 2013, http://www.nsf.org/business/drinking_water_treatment/standards.asp?program=DrinkingWatTre.

42 NSF International, "Consumer Products and Service Listing Search," accessed February 2013, http://www.nsf.org/Certified/dwtu/.

43 National Oceanic and Atmospheric Administration, *About Our Great Lakes: Great Lakes Basin Facts,* accessed March 2013, http://www.glerl.noaa.gov/pr/ourlakes/facts.html.

44 G. Hardin, "The Tragedy of the Commons," *Science* 162 (1968): 1243–48.

45 H. Daly, Center for the Advancement of the Steady State Economy, "Eight Fallacies about Growth," August 2012, accessed April 2013, http://steadystate.org/eight-fallacies-about-growth/.

46 W. Baumol and W. Oates, *The Theory of Environmental Policies* (Cambridge: Cambridge University Press, 1988).

47 J.K. Galbraith, *The Culture of Contentment* (New York: Mariner Books, 1993).

48 See for example: Ellen MacArthur Foundation, "Toward the Circular Economy: Economic and Business Rationale for an Accelerated Transition," Vol. 1: 2012, accessed July 11, 2013, http://www.ellenmacarthurfoundation.org/business/reports/ce2012.

49 T.T. Schug et al., "Designing Endocrine Disruption Out of the Next Generation of Chemicals," *Green Chemistry* (2013). doi: 10.1039/c2gc35055f.

50 Tiered Protocol for Endocrine Disruption (TiPED), *What is TiPED™?*, 2012, accessed June 2013, http://www.tipedinfo.com/tiped_tier/whats-tiped/.

51 "Green Chemistry Success Stories," *Lanxess Webmagazine*, accessed May 2013, http://webmagazine.lanxess.com/green-chemistry/success-stories-abound.html.

52 S. Everts, "Better Living Through Green Chemistry: Pharmaceuticals," *New Scientist*, March 12, 2010, accessed April 2013, http://www.newscientist.com/article/dn18641-better-living-through-green-chemistry-pharmaceuticals.html.

53 Navigant Research, "Electric Vehicles in Europe," May 2013, accessed May 2013, http://www.navigantresearch.com/research/electric-vehicles-in-europe.

Chapter 7

1 L. Parlett, A. Calafat and S. Swan, "Women's Exposure to Phthalates in Relation to Use of Personal Care Products," *Journal of Exposure Science and Environmental Epidemiology* 23 (2013): 197–206.

2 See Table 5 in Chapter 2.

3 N. Klepeis, W. Nelson, J. Robinson, A. Tsang, P. Switzer, J. Behar, S. Hern and W. Engelman, "The National Human Activity Pattern Survey (NHAPS): A Resource for Assessing Exposure to Environmental Pollutants," *Journal of Exposure Analysis and Environmental Epidemiology* 11 (2001): 231–52.

4 C.G. Bornehag and E. Nanberg, "Phthalate Exposure and Asthma in Children," *International Journal of Andrology* 33 (2010): 333–45; C.G. Bornehag, J. Sundell, C. Weschler, T. Sigsgaard, B. Lundgren, M. Hasselgren and L. Hägerhed-Engman, "The Association between Asthma and Allergic Symptoms in Children and Phthalates in House Dust: A Nested Case-Control Study," *Environmental Health Perspectives* 112 (2004): 1393–97; R. Dodson, M. Nishioka, L. Standley, L. Perovich, J. Green Brody and R. Rudel, "Endocrine Disruptors and Asthma-Associated Chemicals in Consumer Products," *Environmental Health Perspectives* 120 (2012): 935–43.

5 K. Ji, Y. Kho, Y. Park and K. Choi, "Influence of a Five-Day Vegetarian Diet on Urinary Levels of Antibiotics and Phthalate Metabolites: A Pilot Study with 'Temple Stay' Participants," *Environmental Research* 110 (2010): 375–82.

6 E. Angell-Anderson, S. Tretli, R. Bierknes, T. Forsen, T. Sorensen, J. Eriksson, L. Rasane and T. Grotmol, "The Association between Nutritional Conditions during WWII and Childhood Anthropometric Variables in Nordic Countries," *Annals of Human Biology* 31 (2004): 342–55.

7 A. Strom and R.A. Jensen, "Mortality from Circulatory Diseases in Norway, 1927–1948," *The Lancet* 1 (1951): 126–29.

8 S. Genuis, "Elimination of Persistent Toxicants from the Human Body," *Human and Experimental Toxicology* 30 (2010): 3–18.

9 Ibid.

10 V. Nadeau, G. Truchon, M. Brochu and R. Tardif, "Effect of Physical Exertion on the Biological Monitoring of Exposure of Various Solvents following Exposure by Inhalation in Human Volunteers: 1. Toluene," *Journal of Occupational and Environmental Hygiene* 3 (2006): 481–89; R. Tardif, V. Nadeau, G. Truchon and M. Brochu, "Effect of Physical Exertion on the Biological Monitoring of Exposure of Various Solvents following Exposure by Inhalation in Human Volunteers: 2. n-hexane," *Journal of Occupational and Environmental Hygiene* 4 (2007): 502–8.

11 V. Nadeau et al., "Effect of Physical Exertion . . . : I. Toluene."

12 D. Kennedy, K. Cooley, T. Einarson and D. Seely, "Objective Assessment of Ionic Footbath: Testing Its Ability to Remove Potentially Toxic Elements from the Body," *Journal of Environmental and Public Health* (2012): 1–13.

13 A. Todd for Environmental Monitoring and Reporting Branch, Ontario Ministry of the Environment, *Changes in Urban Stream Water Pesticides Concentration One Year After a Cosmetics Pesticides Ban*, November 2010.

CREDITS

Grateful acknowledgement is made to the following for the permission to reprint previously published material:

Quotation on page 1 is from "Canary in a Coal Mine," words by Sting, music by Sting and Nigel Gray. © 1980 Universal Music Corporation and Songs of Universal, Inc. Used by permission.

Figure 1 on page 7 is adapted from "Worldwide Trends in DDT Levels in Human Milk" by D. Smith, published in the *International Journal of Epidemiology* 28 (1999): 179-88. Reproduced with permission from Oxford University Press (License No.: 3182491348809).

Figure 2 on page 8 is adapted from "A model to evaluate past exposure to 2,3,7,8-TCDD" by P. Pinsky and F. Lorber, published in the *Journal of Exposure Analysis and Environmental Epidemiology* 8 (1998): 187–206. Reproduced with permission from Nature Publishing Group (License No.: 3180781145797).

Figure 3 on page 9 is adapted from *Risk Management Strategy for Lead*, http://www.hc-sc.gc.ca/ewh-semt/alt_formats/pdf/pubs/contaminants/prms_lead-psgr_plomb/prms_lead-psgr_plomb-eng.pdf, Health Canada, 2013. Reproduced with permission from the Minister of

Public Works and Services Canada, 2013 (File No.: 2013-36969).

Table 2 on pages 28–9 is adapted from Adria Vasil's *Ecoholic Body* (Vintage Canada, 2012). Used by permission.

Quotation on page 87 is from "Rehab—Remix" words by Amy Winehouse and Jay-Z, music by Mark Ronson and Jay-Z © 2007 Islands Records and Universal Music Corporation and Songs of Universal, Inc. Used by permission, provided through the courtesy of Hal Leonard Corporation.

Quotation on page 148 is from "Smoke Gets in Your Eyes," *Mad Men*, Episode 1, Season 1, July 19, 2007. *Quote taken from "Mad Men" provided through the courtesy of Lionsgate.*

Figure 11 on page 155 is adapted from "The National Human Activity Pattern Survey (NHAPS): A Resource for Assessing Exposure to Environmental Pollutants" by N. Klepeis, W. Nelson, J. Robinson, A. Tsang, P. Switzer, J. Behar, S. Hern and W. Engelman, published in the *Journal of Exposure Analysis and Environmental Epidemiology* 11 (2001): 231–52. Reproduced with permission from Nature Publishing Group (License No.: 3171361345710).

Figure 14 on page 214 is reproduced with permission from the Ellen MacArthur Foundation.

INDEX

Italicized page numbers refer to pages with tables or figures

BRUCE LOURIE is a leading environmental thinker, writer, and speaker. He is president of the Ivey Foundation and is a director of several environmental and energy organizations in Canada and the United States.

RICK SMITH is a prominent Canadian author and environmentalist. He is executive director of the Broadbent Institute and was the executive director of Environmental Defence for almost ten years.

They both live in Toronto.